Bargaining with the Machine

BARGAINING WITH THE MACHINE

Technology, Surveillance, and the Social Contract

Robert M. Pallitto

University Press of Kansas

Published by the University Press of Kansas (Lawrence, Kansas 66045), which was organized by the Kansas Board of Regents and is operated and funded by Emporia State University, Fort Hays State University, Kansas State University, Pittsburg State University, the University of Kansas, and Wichita State University.

Library of Congress Cataloging-in-Publication Data

Names: Pallitto, Robert M., 1964– author.
Title: Bargaining with the machine : technology, surveillance, and the social contract / Robert M. Pallitto.
Description: Lawrence : University Press of Kansas, 2020. | Includes bibliographical references and index.
Identifiers: LCCN 2020007580
ISBN 9780700629848 (cloth)
ISBN 9780700629855 (paperback)
ISBN 9780700629862 (epub)
Subjects: LCSH: Technology—Social aspects. | Electronic surveillance—Social aspects. | Social contract.
Classification: LCC T14.5 .P35 2020 | DDC 303.48/3—dc23
LC record available at https://lccn.loc.gov/2020007580.

British Library Cataloguing-in-Publication Data is available.

Printed in the United States of America

10 9 8 7 6 5 4 3 2 1

The paper used in this publication is recycled and contains 30 percent postconsumer waste. It is acid free and meets the minimum requirements of the American National Standard for Permanence of Paper for Printed Library Materials Z39.48–1992.

To Ana Meléndez

and

Ellen and Roger Miller

CONTENTS

ACKNOWLEDGMENTS

First and foremost, I owe my deepest thanks to David Congdon at the University Press of Kansas for sticking with this project through its many twists and turns, and for his keen insights that kept the work on track and made it better. I was fortunate to benefit from the existence of the Surveillance Studies Network in a number of ways: for insights gleaned from conferences and articles, for constructive feedback on many of the ideas contained in this book, and for the network's high level of scholarly inquiry that challenges and stimulates. In particular, I am grateful to Torin Monahan for his insightful advice, his encouragement, and the inspiration he provides with his own scholarship.

I have presented and published papers related to the themes of this book in a number of different venues, and the feedback I received in the process was most valuable. Those venues include conferences convened by the Surveillance Studies Network, the Urban Affairs Association, the American Political Science Association, the Northeast Political Science Association, and the Seton Hall Math Department's Cybersecurity Workshop. I am grateful to *Surveillance and Society*, *First Monday*, and the *Journal of American Culture* for the opportunity to express my thoughts about technology, politics, and surveillance in print.

I have benefited from the advice and encouragement of several scholars whose work I admire, especially Patricia Williams, Gray Brechin, Torin Monahan, Kent Worcester, John Feffer, Gary Marx, and Frank Pasquale. The anonymous reviewers of this manuscript also have my thanks for their constructive criticism that enriched the final product.

Here at Seton Hall University, I received help and support from a number of friends and colleagues as I worked to bring the book to completion. Lisa DeLuca has been a generous and critical reader as well as a constant source of newly published research, especially about smart cities. In my home Department of Political Science and Public Affairs, Jeff Togman, JoRenee Formicola, Mike Taylor, and Geoff Upton reviewed drafts and/or provided critical feedback. Antonella Fegan contributed much-needed research support on trusted traveler programs. I am fortunate to work with a number of supportive Seton Hall colleagues. I am particularly indebted to Jonathan Farina, King Mott, and Janine Buckner, for reasons that will be immediately apparent to them.

As always, family and friends have been behind me with unwavering support and confidence as I wrote this book. Warmest thanks to Kathryn Motoviloff, Ellen and Roger and Andrew Miller, Ana Meléndez, Christina Pallitto, Ben Smith, John and Kerry Motoviloff, Bob and Gwyn Murray, Emily Bregman, Doris Masse, Gerard Byrne, Jaya Christensen, Jerry and Karen Wahl, Frank Magaletta, and John and Daniel and Isabel Meléndez. Finally, my most profound gratitude to Laura Meléndez, who for more than three decades now has been an elemental force in my life, like the moon causing the tides to move.

I hope that my imperfect memory has not led me to forget anyone who has helped me over the years that I have been working on this project. All errors, as the refrain goes, are solely my responsibility.

CHAPTER ONE

Introduction: Irresistible Bargains

> Whatever remains unsatisfied in them through the order which takes from them without giving in exchange what it promises, only burned with impatience for their gaoler to remember them, and at last offer them stones in his left hand for the hunger from which he withholds bread in his right.
>
> —Theodor Adorno, *Minima Moralia*

Not long ago, a Swedish company began implanting microchips in its employees.[1] Each chip was injected by syringe into the flesh separating thumb and index finger. Once in place, it would allow wearers to open doors, operate printers, and buy food. Employees who consented to the procedure explained their willingness to permit the injection. "I want to be part of the future," said one. Another judged the process to be less invasive than other implantations, such as pacemakers. The CEO of the company (who also carried an implant) cited convenience as the principal benefit. While their responses varied, each of the participants had agreed voluntarily to engage with a technological process that enables communication and commerce on the one hand, but pierces the skin and forces the body to "emit signs" on the other.[2]

How often do individuals choose to engage with technology, whether in terms as graphic, literal, and invasive as the chip injection, or in more subtle ways? Quite often, as it turns out. Technological artifacts like the Swedish microchip that create a "cyborg"—a "hybrid of machine and organism"—abound, and they are not particularly new.[3] The hearing aid, pacemaker, stent, and prosthetic limb were in use long before the subcutaneous microchip. In fact, all four of those devices also predate the wearable insulin pump and wearable heart monitor, both of which also fuse machine with organism. And of course, not all of today's technological innovations involve cyborgs. Many of the technological devices and processes that have become part of our world do *not* inhabit our physical bodies. Cell phones, computerized toll collectors, biometric scanners, and closed-circuit television cameras are all extraneous to our bodies, but they affect us nonetheless in ways that we anticipate as well as ways that we do not fully understand. In other words, people make choices to engage with many forms of technology in everyday life, and some of those forms are invasive in obvious ways while others have more obscure or insidious effects. In all cases, it is possible to examine the choices that bring us into contact, or engagement, with technology, and even to ask sometimes whether we have any choice at all.

How do people perceive the choices they make when they use social media or EZ Pass, or when they consent to be microchipped? Why do they consent when there are costs involved? In this book I explore the experience of living in a world where technology is central to so much of what we do. Most people would freely acknowledge the important role of technology in their lives and would likely admit that they continually expect certain tasks and activities to be made easier or more efficient by a new technological innovation. We have become increasingly focused, in this "Age of Machinery,"[4] on instruments that make our lives easier, and in fact we are often motivated by ease, or convenience, or efficiency, when we adopt a new device or consent to an intrusion. This notion of *ease* is a catch-all category that takes in everything from life-saving medical procedures to pool-cleaning robots. At times, being "part of the future," as one of the Swedish employees averred, is an added inducement to engage with technology. It means having the opportunity to utilize a new convenience first, before everyone else does, to stand amazed, and to amaze others, by what is now possible.

Technological Engagement as Bargaining

I propose that when people decide whether to engage with a technological process or artifact, they appear to be engaging in a form of bargaining. They compare benefits to be gained with costs to be incurred and then decide whether a particular choice or action is warranted. That choice can be about consenting to an intrusion (e.g., a search or a drug test), or it can involve the use of a tool, such as GPS navigation or an online dating site. This focus on cognitive deliberative processes at the individual level makes it possible to identify a person's motivation in deciding to utilize or engage with a particular artifact and also allows some form of evaluation of those decisions in terms of expected versus actual gains and losses. Through the analytical frame of bargaining, we gain a stronger understanding of how subjects are situated as those choices confront them; we see more clearly how their life-chances are affected; we gain insights into subjects' self-understanding. The individual making the decision (or in technical terms, the bargaining subject) mirrors the subject imagined in classical economic theory: a rational, calculating actor who seeks to maximize self-interest. This figure has appeared in modern social and economic thought at least since Adam Smith celebrated the human "propensity to truck, barter and exchange."[5] It is utilized in many forms of inquiry and endeavor and has generated habits of thought spanning domains from tort law to product marketing. So it is a familiar figure, and one that is often used not only to evaluate a choice after the fact but also to predict in advance what a person—a *rational* person—will do, in advance.[6] In what follows, I will be advocating a more critical stance toward the construct of bargaining, but its prominence in social, political, and economic thought is undeniable. This prominence is all the more reason to look closely at it.

Toward that end, the following questions can serve to structure an inquiry concerning technological choice as a form of bargaining—and to show that in many instances, an individual's ability to bargain freely is limited.

Who is proposing the bargain? Is it the state, as when individuals are asked to submit to invasive airport searches in order to be safe from terror attacks? Alternatively, private actors offer to bargain, for example, "This call may be monitored (surrender your privacy) in order to serve you better (you'll get something in return)." Additionally, bargains can be nested inside one another, so that the state proposes a bargain through legal frameworks that diminish privacy (at airports, for example), and nested within that bargain is the one a passenger strikes with a Transportation Security Administration official at the airport: bargaining occurs at both levels. This "nesting" phenomenon, whereby one bargaining relationship grows out of another, can help to show how a one-on-one, state actor–individual bargain (TSA agent searching a passenger) is actually a smaller version of the higher-level relationship with the state in which the possibility of bargaining is limited or nonexistent. The bounds of what is permissible or possible in a face-to-face interaction is shaped by the state, which put the individuals face-to-face to begin with. At the higher level (where the individual confronts the state), the bargaining partner is so vast and impersonal that there does not seem to be a bargain at all—at least not in the sense of two sides seeking mutual advantage through negotiation. In such cases, thinking about the framework of bargaining serves to show that no meaningful form of bargaining can occur in some conditions. Just as some bargains nest inside each other, bargains generate micro-bargains—as, for instance, when the decision to utilize a form of technology subsequently poses smaller questions. One might decide that using an Internet search engine is worth the risk that their activity may be tracked, but then each decision to utilize a notorious search term such as "terrorism" or "bomb-making" requires the consideration of what notice that term might trigger.[7]

What is the balance of bargaining power? Individuals are almost never bargaining at arm's length, with equal bargaining power. The state obviously has greater resources, and even in cases where a bargain is struck with a powerful *private* actor such as a multinational corporation, freedom to refuse or to demand better terms can be a fiction that is parasitic upon the (neo)liberal fiction of freedom of contract. We say that individuals are free to make any contract they like, and yet in practice one side often has much more to lose. Imagine, here, a factory owner offering low wages to a worker. The worker is "free" to refuse only if better prospects are out there. And the worker has much more to lose than the owner by refusing. By asking about the balance of bargaining power we can often discover that no meaningful choice exists, and that what looks like a bargain is actually a situation where an individual concludes there is no alternative. Questioning the balance of power in bargaining recalls Jurgen Habermas's use of the "ideal speech situation" in discourse ethics: two speakers of equal power and status generate an agreement, through deliberation, that is acceptable to each of them.[8] In both cases—speech and bargaining—the ideal type can help to diagnose what is missing in real-life encounters between subjects. As Gary Marx puts it, "To be meaningful choice should imply genuine alternatives and refusal costs that are not wildly exorbitant, absent that we have trickery, double-talk, and the frequently spoiled fruit of inequitable relationships."[9]

Moreover, the relative power and efficacy of different individuals and groups can also inflect the bargaining relationship. Historian Sarah Igo has noted that vulnerable individuals in a society may face a bargain with the state that is harder to resist for them than it is for others.[10] She uses the example of public benefits such as welfare grants. In exchange for receipt of such benefits, applicants and recipients are required to submit to various intrusive measures that diminish their privacy: informational privacy as well as spatial privacy and intimate details of sexual history. In specific terms, data on work history, family history, and intimate relationship history are compiled, tracked, and shared by agencies. In exchange for welfare benefits in particular, mothers are required to cooperate in establishing the paternity of their children, and that process involves public discussion of sexual history as well as the identification of name and whereabouts of putative fathers.[11] Needless to say, the paternity process generates some risk of retaliation by fathers who did not want to acknowledge their children, as well as indignity when women are compelled to share the most intimate details of their lives with a caseworker who is a total stranger (often, more than one total stranger). Igo traces this bargain back to the passage of the 1935 Social Security Act, which established the practice of identifying and tracking individuals by Social Security number. In exchange for this government assistance that helped meet a range of needs including food, clothing, and shelter, individuals had to become "known citizens," in Igo's terms. For many—during the Great Depression years when Social Security began, and continuing through the present—this was a bargain they simply could not refuse, and thus not a bargain at all.[12]

A final related question here would be, "How familiar/recognized is a given bargain?" This question would help to illuminate the social significance of the choice being offered. In the context of a given set of social conditions, how, if at all, would the choice be recognized or understood by one's peers in that setting?

What is surrendered? What is gained? Sometimes it is clear what an individual gives up in deciding to engage with a technological artifact that promises information storage or social networking or the like. When employees agree to unannounced workplace searches, to take another example, they are clearly losing an expectation of privacy that they would otherwise have (in exchange for keeping their jobs). At other times, individuals may not know the full extent of what they will surrender. Personal data that I provide in a retail transaction might be shared via lists for marketing purposes or assembled into data profiles for the same reason. And when that data is utilized to address consumers by targeted marketing, those consumers might not know how information used to target them was obtained, what was learned secretly, or what calculus produced the decision to address them rather than someone else. In other words, we surrender information at one point in time that leads in turn to "downstream" consequences that we don't connect to the original act of surrender. In fact, we might never know all of the uses to which the surrendered data will be put.

Just as we don't fully know what is given up in a bargain, what we gain (i.e., the benefits of the bargain) are often unclear or uncertain as well. Consider the tradeoff that has been proposed innumerable times since the terror attacks of 2001: submit to intrusive forms of scrutiny by government at the airport, on the telephone, and even at home (sacrifice liberty) so that the government can prevent another attack (in exchange for security). In reality, we can never know that we have been made safe by our sacrifice of liberty because safety one day could be followed by an attack the next day. As Torin Monahan explains, the liberty/security trade-off does not always bring greater security—and in fact, the government often explicitly disclaims any ability to keep us safe even while demanding new and greater sacrifices of liberty.[13] This is one example of lack of clarity or certainty regarding gains.

What are the actual versus perceived gains and losses (what is hidden)? This question lies at the heart of the bargaining inquiry. In one sense, it is a predictive question, as when it is applied to the liberty/security tradeoff. Will airport searches prevent hijackings? It is difficult to answer with certainty, but one could at least utilize statistical and anecdotal evidence to make a judgment. Gary Marx suggests that in the ideal bargaining relationship, "Individuals fully understand not only what they will be receiving but also what they are giving away, how it will be used and protected, and what potential risks and secondary uses there might be."[14]

But of course, most bargaining relationships fall short of this ideal, frequently because there is hidden information that is available to one party and not the other. In the aforementioned targeted marketing example, customers receive product offers, which they are of course free to accept or decline. What they may not know, however, is how the marketing information is obtained and why a particular customer has been chosen to receive it. The retail giant Target allegedly utilized targeted marketing to reach pregnant women on the theory that they would be more receptive to certain sales pitches as compared to nonpregnant women, and men.[15] Company statisticians were able to predict through purchasing patterns which customers were likely to be pregnant. Customers had not told anyone at Target that they were pregnant, of course: the conclusion was reached through aggregating purchase data. Such background information revealing how she was chosen would be useful to the customer, and might even influence the decision to accept or reject the bargain. If she knew that the retailer had determined surreptitiously that she was pregnant, and then targeted her for product offers on that basis, she might decline the offers based on that knowledge. This is another way in which a bargaining frame illustrates the lack of a meaningful bargaining relationship: hidden information makes a bad bargain (or no bargain). With regard to state (as opposed to private) actors concealing effects of information disclosure, Cory Doctorow has this to say: "If online oversharing is a public health problem, then the state's decision to harness it for its own purposes means that huge, powerful forces within government will come to depend on oversharing. It will be vital to their jobs—their paypackets will literally depend on your inability to gauge the appropriateness of your online disclosure."[16]

When information has been hidden, including information about why data is being gathered or how it will be disclosed in the future, that is a reason to say that a particular bargain is a bad one. Such a scenario is analogous to fraud in contract formation. Fraud voids a contract because the victim of fraudulent inducement did not make a conscious and informed decision; their assent was premised on facts that were deliberately misrepresented. Similarly, a person who agrees to engage with a process or device based on incomplete information with vital facts hidden did not make an informed decision. Their consent was obtained in a way that calls its validity into question. The bargains people make involving technology often come at such steep and poorly understood costs that they call into question whether such bargains were freely chosen at all.

More difficult cases arise where an individual explicitly avows that the bargain was struck for a particular reason. "I know I'm surrendering personal information to obtain an expedited border crossing permit," one might say, "but I'm doing it willingly because I can travel much faster and more conveniently." Phrasing the bargain that way, we make it difficult to criticize. We might be tempted to say, "It's not worth it," but in fact the individual has already decided that it is. People surrender control in exchange for convenience. They surrender control to facilitate lifestyle. At the end of the bargaining inquiry, when all the questions I list here have been explored, we will sometimes reach a point where no normative judgment "sticks" to criticize a bargain struck by an individual. If the bargainer is getting what is sought (convenience, privilege, desired consumer goods), a critic is hard-pressed to call the bargain "bad," "false," or "wrong" for the bargainer. Let me be clear, though: there is still more to say even if normative criticism is problematic. For one thing, the consequences of the bargain can be drawn out and described. Consider the case of a home buyer deciding to live in a gated community. The buyer may be well aware that the home purchase will entail a loss of privacy within the walls of the community, as well as a certain isolation from the world outside. Nonetheless, the buyer deems those losses worth the bargain in exchange for living in a controlled environment where visitors are screened. It is difficult to find a common metric whereby the components of the trade-off can be measured, and so a critic might conclude that it is not possible to evaluate the bargain normatively and call it good or bad, at least if there is no universal set of interests or ends against which to judge a bargain. The doctrine of freedom of contract assumes that individuals are free to choose their own ends provided those ends are not illegal. Nonetheless, even here, it is possible to trace out the effects of the trade-off and then consider those effects in future policy-making. The walling off of private spaces by the designing of gated residential communities contributes to civic decay, social stratification, and general suspicion. These effects can be documented even though we may be unable to say that the resident made a bad bargain in terms of individual self-interest. Social disapprobation can—and sometimes should—be taken into account by a person making a bargaining decision. The bargains people make, and how they feel about them, can be shaped by social norms and judgments. Going against collective judgments and expectations can be difficult, and even traumatic.

The Paradox of Bargaining in a World of Limited Agency

There is a paradox involved when we envision bargaining in a complex and impersonal world, where social forces and conditions shape the possibilities for taking action. Liberal political thought ascribes great importance to agency and free choice while postmodern theory questions the possibility of such freedom. This is a paradoxical tension in the sense that invitations to act and choose in many contexts abound even though the limits of freedom are often obvious. Claims about individual freedom, and laments over powerful external forces that diminish such freedom, both feature prominently in the background thought of our contemporary world. Consumers are addressed as sophisticated people who design their lives according to highly specific preferences in food, clothing, travel, entertainment, and relationships. Voters now enjoy a volume of information available to them that is unprecedented in the history of democratic government. These interactions suggest that individuals are in control of the conditions of their lives.

Assurances that this is so—that we *are* in control—are abundant. So many subject positions presented in contemporary social life seem to suggest the capacity to "choose freely." Consider, for example:

- *Consumer* (In fact, "Choose freely" is an actual slogan used by the Coca-Cola Company in urging people to use its advanced soda machine, "Freestyle," which offers "an unprecedented array of drink choices in a fun, interactive format."[17])
- *Party to the social contract* (Consent is the legitimating element in modern democratic government: we constitute ourselves as a republic by agreeing to be bound by its laws. Consent is the foundation of the

polity, whether we draw on the words of early social contract theorists or the current popular discourse around legislative representation.)
- *Voter* (Voting is a civic duty promoted in public service announcements and urged on us as a social norm. Voters expend the effort to enact an informed choice in exchange for an electoral result they want and a sense of fulfilling a duty.)
- *Modern individual who freely chooses relationships with others* (The Supreme Court has said that personhood is a vital aspect of liberty that develops through our choice of personal associations throughout the course of our lives.[18])

Each of these roles, or *subject positions,* contemplates a person who can and does make free choices. Advertisers address consumers, and candidates address voters, expecting that an individual choice will be made and hoping to influence that choice to vote or to buy.

At the same time, social forces can make those choices seem illusory. Consumers are *not* free to shape the terms of the bargain when purchasing goods from a large commercial entity such as Amazon. As will be explored more fully later in this book, Amazon functions increasingly as not only the bargaining partner in many transactions, but also as the *judge* of the bargain as well, as Frank Pasquale shows convincingly through his analysis of the "functional sovereignty" of "platform capitalism."[19] Firms like Amazon attain this "functional sovereignty" through their control of platforms so that they impact people's behavior and life-chances similarly to the way governments who control physical space enjoy "territorial sovereignty." They are "able to exert regulatory control over the terms on which others can sell goods and services" by such means as offering rating services for consumers seeking a bargaining partner, and dispute resolution between buyers and sellers.[20] Thus, Amazon is at once a bargaining partner (when we buy through them) and the setter of rules by which that bargaining takes place.

Similarly, voters often receive incomplete information about political candidates as the candidate images are crafted and shaped by media outlets and by the candidates' campaign strategists. As a result, the freely choosing subject imagined in the aforementioned bargaining situations may not correspond to people's actual lived experience. Moreover, it is evident that people face powerful obstacles to free choice, such as ever-more-efficient surveillance, precisely targeted marketing, and media centralization. These constraints are not merely topics of sophisticated theorization but also the constitutive facts of our lived experience. Online shoppers leave a purchasing history behind that allows marketers a fine-grained picture of buying habits and preferences. With knowledge of that history, sellers can target buyers with unerring precision, shaping their habits even as they track them. News information and advertisements come at us via the Internet at once, in a carefully created package. As consumers of goods, services, and information, we are nudged subtly and continuously toward some choices and away from others. Some commentators have even claimed that people are so well accustomed to their loss of privacy, for example, that they no longer care about the loss.[21]

The tension between claims of agency and awareness of its limits is the site where we experience our social lives. The relationship is not stable or precise: we cannot pinpoint a fixed spot on a spectrum between complete freedom to act, on one end, and powerlessness in the face of external control, on the other, and say that it represents the extent of agency that people enjoy. In fact—to cite the words of one commentator—there is a "messier" relationship between those "who employ the technologies of surveillance in an effort to subordinate individuals to societally constructed identities that ultimately sustain entrenched an increasingly postdemocratic political order and those, on the other hand, who in their collective/collaborative endeavors as performance practitioners strive toward a sense of community that might effectively counteract the interpellation and social sorting that surveillance technology facilitates."[22]

This formulation describes the "messiness" of the social/political orders in which we live and act. Surveillance technologies limit possibilities for action and choice.[23] Describing the order of things as "postdemocratic" suggests that chances for popular input and control over our governance are closed off and even exhausted. In a sense, this term refers to a world in which the time for democratic practice has passed. But practicing "performance" can be a response to our encounters with postdemocratic phenomena.[24] Performance takes place in traditional art spaces but also on the streets, in protest activity. And both of these kinds of performances are often *collective,* so that agency can be realized in collaboration when it might have been thwarted for an individual.

In view of this complex and contested relationship, freedom and agency in the surveillance society is a tension to be mapped and explored even if it is paradoxical. We often act *as if* we had complete freedom to choose even as we confront continual reminders of the limits of freedom. This book will attempt to describe the ways in which that complex, conflicted, and even paradoxical experience arises and presents itself to us.

Machines as a Threat to Bargaining Freely ("Everywhere It Is Machines . . .")

Sometimes, the bargaining partner, or the entity with which a person must contend when seeking to make a technological bargain, seems so vast, powerful, and impersonal that we might conclude no bargaining can take place. In such instances, the bargaining paradigm is useful for showing us that there is *no room* for bargaining. So many representations of machines (or the *machine*) have been imagined through the course of modernity that this Other has become easy to visualize. Think of Simone Weil's "social machine" that crushes souls and bodies.[25] The cyborg itself. Gilles Deleuze's "machines being driven by other machines."[26] The Mechanical Turk, an eighteenth-century chess-playing "machine" with a little human inside, resurrected first by Walter Benjamin and then more recently by Amazon as a crowd-sourcing platform. The machine is a formidable bargaining partner—so much so that it can be impossible to stand against it and seek one's own advantage.[27]

Some of the instances that look like bargaining are in fact cases where the outcome is determined in advance, despite what the human agent may think. In addition to the other questions to be asked when evaluating a given social bargain involving technology, it is often necessary to ask whether any bargaining did or could take place. For this inquiry the figure of the machine, which looms large in so many dystopian critiques of (post)modern society, is suitable to introduce. It will appear in various forms throughout this book.

Outline of the Book

The argument here will proceed by examining different kinds of human-technology encounters as bargains and considering to what extent engagement with technology in each type of case is an irresistible bargain and thus no bargain at all. Chapter 2 explores the related notions of convenience and efficiency, both of which can make a given technological artifact appear irresistible. Many digital media-related products are offered free to the user as the provider is not seeking immediate payment but rather collecting revenue from advertising and secondary transactions with data brokers. Further, online transactions are made so simple for the consumer that no more than a touch of the finger is required to complete them. But the ramifications of such decisions are poorly understood. The convenience of a cell phone app or a search engine obscures the ways in which the engagement with that product causes personal information to flow away from the individuals and out of their control. One is not free to bargain for ways to stop or slow that information flow. Even if a consumer bothers to read the privacy policy for Facebook or Google (and many of us do not), there is no room for objecting, changing, or renegotiating terms. Bargaining is nowhere near equal. As a result, convenience, though it appears irresistible, comes at a cost.

Efficiency is closely related to convenience in discussions of technological choices. To say that a person is motivated by convenience when taking a certain action, we are focusing on ease of effort. It is more convenient to shop online because we can do it without moving off the couch. Using the word *efficiency,* in contrast, to describe the same choice suggests some kinds of savings, a decrease in cost. Of course, not only monetary cost but also time and opportunity costs are relevant in this calculation. But the focus in efficiency terms is on the savings, while an assessment of convenience has more to do with ease of effort.

Two types of technologies in particular exemplify the lure of convenience and efficiency and the corresponding costs. Accelerated air travel programs and customer loyalty programs can seem irresistible by virtue of the savings in time and money that they promise. TSA PreCheck in the United States and Registered Traveller in the UK enable travelers to more efficiently move through airport security and passport control in exchange for a fee and the submission, in advance, of personal data that allows the government to assess risk level.

Travelers who qualify as sufficiently low risk are granted the status of preferred traveler, trading off privacy for efficiency. These programs serve as an example of what Nikolas Rose calls "circuits of inclusion," which he defines as "recurrent switch points to be passed in order to access the benefits of liberty."[28] Not all individuals can move within these circuits, and the ones who are able to do so have traded, at least in the preferred traveler case, control of personal data in order to gain more efficient movement. In place of regular, time-consuming, low-tech searches, they have been searched one time in a more pervasive and fine-grained manner. As part of this project, I applied for (and received) TSA PreCheck status, and I report here on the costs and benefits as well as the symbolic significance of my "special status" as a traveler. The status differentiation I experienced vis-à-vis other travelers was notable and cannot be discounted in considering the possible motivations one might have for entering into this kind of bargain.

Customer loyalty cards and other preferred customer programs are similar to air travel programs in that all of them promise a benefit in efficiency and/or convenience to a person who grants access to personal data. In the loyalty card example, one's purchasing history is laid bare to the retailer (or gym, or cinema, etc.) in exchange for discounts, special offers, and other preferential treatment. Of course, one value of the loyalty card for the retailer—at least for bigger retailers—is that it allows data collection that can be turned into profit when the data is sold to others. Here, again, convenience and efficiency are only part of the allure of the loyalty card. There are also affective rewards awaiting the "loyal customer" who performs what Pramod Nayar has termed "consumer citizenship."[29] Card users learn ways of behaving that make them valuable members of a community shaped and defined by commerce, by consumption. Instead of voting or signing petitions or running for local office, consumer citizens belong because they purchase goods and services in a certain way.

Chapter 3 addresses the ubiquity and obscurity of technological bargains. Certain forms of technology, such as online shopping and social networking, are present all around us, confronted at every turn, so that functioning without using them is nearly an insurmountable challenge. As people become accustomed to such activities, it becomes more difficult to find settings where they are not in use. Amazon, for example, has begun using cashier-less stores, which are nothing more than a shell, a physical space where transactions take place without a transactional partner. Data collection occurs even though no one is there to collect it. And so data collection points are found throughout the environment; they *are* the environment. David Murakami Wood terms this pervasive and taken-for-granted watching "ambient surveillance."[30] Ambient surveillance has a self-reinforcing effect in that people demand emerging technologies in new settings after becoming accustomed to seeing and using those technologies. Think of cell phones versus pay phones as one example. If someone wanted to eschew cell phone use and rely on pay phones in public phone booths (when was the last time you saw one?), would that be feasible?

In the chapter, ubiquitous bargains will be explored primarily in connection with the platforms offered by companies like Amazon. The platform offers such a wide range of products and services that consumers encounter it when they shop for almost anything. There have been other ubiquitous technologies structuring the life-world at other points in time, of course, but platform capitalism has ushered in new ways of drawing in consumers that make it a particularly useful grounding for studying ubiquitous technologies that can become irresistible at times.

Just as the ubiquity of certain technological forms and processes makes them hard to avoid and therefore hard to resist, so does the obscurity of certain technological bargains draw us inexorably into their operation when we do not know whether or how they are operating. If personal information circulates as valuable data sought by data brokers and then by marketers, we would predictably want to learn how we lost control over that data in the first place, but often it is impossible to back-trace the disclosures that led to its publication and transfer.

Large-scale data collection and the activities of data brokers constitute a vivid example of obscurity in technological bargains. Data is collected, repackaged, and shared with marketers in ways that make it difficult for individuals to pinpoint the sources of sharing, much less to control the personal information that is circulating. Moreover, data brokers create demographic types that operate as "data doubles," composite types that stand in for the individuals from whom the data was collected. One particular harm that results from this process of data collection, aggregation, and sharing consists in the substitution of constructed (stereo)types from freely acting and choosing individuals. The chapter will explore the harms to the self resulting from the activities of data brokers and the creation of data doubles and shadow selves.

Chapter 4, which explores the interconnectedness of the world we live in (the life-world) as an inducement to technological engagement, takes as its grounding context the Internet of Things (IoT) and smart city initiatives. Jathan Sadowski and Frank Pasquale describe the IoT as "sensors and computation embedded within physical objects that then connect, communicate, and/or transmit information with or between each other through the Internet."[31] Here are some examples:

- A cardiac patient wears a monitor that reports heart function to the physician's office.
- A runner uses a fitness watch to measure performance.
- A webcam monitors an eagle's nest so Internet viewers can follow the hatching and growth of nestlings.

The Internet technologies spawned by the IoT lead quickly to the creation of public or quasi-public spaces that are shaped and arranged around smart technologies. As a result, "It no longer makes sense to think of 'the Internet' as a thing that one accesses via a computer. Not when the city itself is reimagined and reconstructed as a platform for and node within networked information communication technologies."[32] IoT technologies—connected, proliferated, and centralized—bring into being the "smart city." The networked life-world can make each engagement with technology seem irresistible as they are encountered constantly, one after another. Smart technologies managing traffic flow, trash collection, crowd control, building entry, and so on hardly allow time to refuse engagement before the next artifact is encountered. The concept of the rhizome will be introduced to describe the development of power networks in the smart city. In nature a rhizome is a horizontally growing root system that branches out with ever-greater complexity. Gilles Deleuze and Felix Guattari applied the concept to social differentiation in their famous work, *A Thousand Plateaus*. Their use of the term has been described as a "de-centered network" in contrast to the "top-down" arrangement of power that is envisioned by other theorists. Commentators have conceptualized the rhizome as "more and more de-regulated flows of energy and matter, ideas and actions."[33] Its endless productive branching effectively characterizes a process in the social field that is not under the control or direction of an agent or author. Just as new roots or nodes beget even newer ones, so do individual social actors reproduce and transmit energy, ideas, and the like.

After depicting the aspects of the smart city and suggesting the difficulty in resisting them, the chapter will interrogate that apparent irresistibility. A techno-evangelist would suggest that the deployment of smart technologies is the only way to govern public spaces and that the tide of progress in installing smart infrastructures is inevitable. But it is worth asking whether neoliberal technocratic ideology could be challenged and replaced with an entirely different or at least more nuanced outlook on governance. The "corporate contract" by which smart technologies are used and installed benefits the government and the private firms offering the technology, but can shortchange the city's residents. A critical evaluation of the assumptions and the flow of benefits associated with smart cities will focus the meaning of irresistibility in this context.

Chapter 5 describes the avenues of resistance that have developed in response to technologies of surveillance: countersurveillance, encryption, cryptocurrency, cash transactions, and surveillance art. In keeping with the book's general approach of outlining both openings for agency and structural constraints, the chapter will also note the limits and paradoxes associated with each potential form of resistance.

Chapter 6 will assess the state of the social contract (a central tenet of modern political thought) given the proliferation of technologies described in the preceding chapters that make the engagement with those technologies seem on its face to be irresistible. The place of the social contract in the normative foundations of democratic government allows us a basis for assessing the social and political harms resulting from seemingly irresistible technologies, and it sutures the break *between the freedom to bargain that we imagine we have and the structures that abridge and sometimes eliminate that assumed freedom.* For the very form of our government rests on the notion of consent. It is consent that makes the political order legitimate. When individuals make other bargains to obtain particular ends, those actions can be evaluated, in normative terms, against their effects on the social contract. Even when an individual gets what was bargained for, the bargain can be criticized for its resulting public harms—harms that are external to the individual bargainer but not to the other parties to the social contract. In contemporary societies, matters are complicated by the simultaneous memberships that subjects hold in a society of citizens and a society of consumers. Those memberships are not easily disentangled.

Regardless of the difficulty, it has never been more important to make this assessment than it is now, when crises of global insecurity regarding resources, violence, and the ecological state of the planet itself seem to worsen and proliferate daily. In view of those conditions, the "bargaining" analysis of technology presented in this book allows us to do two things:

1. To show that some bargains are so exploitative and unequally derived that they are not bargains at all.
2. To show that even knowingly and voluntarily chosen bargains that are good for an individual are nonetheless bad for the larger society.

My aim in this book is to provide a better understanding of what we do when we make bargains to utilize forms of technology, and how these choices are at once intentional and voluntary but also conditioned, structured, and less than fully understood. Critique is still possible—not to say necessary—despite the larger forces that limit individual choice. Lack of individual freedom does not preclude critique of that freedom.

CHAPTER TWO

Technologies of Convenience and Efficiency

> Trust takes a backseat to convenience for most.
>
> —Survey respondent, Pew Research Center

> We know that people are wildly uncomfortable with the amount of information that, e.g., Google, has about them, but it does not stop them from using Google. People need to live their lives and they will use the services they find necessary.
>
> —Survey respondent, Pew Research Center

Introduction: An Irresistible Surveillance Society?

Promises of convenience and efficiency become touchstones of irresistible bargains, inducing people to embrace various forms of technology despite the privacy risks and other social harms that those technologies pose. As a result of consumers' endless search for greater convenience and efficiency, data collection and other surveillance practices have become more pervasive. Taking stock of the growing scope and sophistication of data collection, one might think that not only the individual convenience-driven bargains themselves, but also an entire society structured by data collection practices, is irresistible. This irresistibility has been characterized as follows: "A surveillance society is inevitable and irreversible. More interestingly, I believe a surveillance society will also prove to be irresistible. This movement is not only being driven by governments; it is being driven primarily by consumers—you and me—as we eagerly adopt ever-increasing numbers of irresistible goods and services, often not knowing what personal information is being collected, or how it may end up being used."[1]

Here, the assumption seems to be that consumers drive this process. They seek convenience/efficiency as they "eagerly adopt ever-increasing numbers of irresistible goods and services," and so the collection of personal information grows and branches, spawning new collection techniques and invading new domains of social life. In the formulation cited above, the consuming public bears the blame for an irresistible surveillance society, as their preferences for convenience and efficiency drive the proliferation of new data collection and surveillance initiatives. More surveillance is irresistible, this view suggests, and it's our fault.

But the consumer is only part of the equation. Consumer demand does not form in a vacuum, but rather in accordance with the goods and services being offered in a given market. Private vendors make deliberate choices when they offer a product to the public: they decide the terms of the offer and what they will require in return.[2] Digital media-related products are free to the consumer, in many cases, since the offeror is not seeking immediate payment from the consumer.[3] Google mail is an example of such a product that offers *double* convenience seemingly without cost. It is easy (convenient) to obtain, and it simplifies (makes convenient) subsequent electronic communication. In line with the imperatives of profit, however, the offeror needs to obtain a benefit when transacting with a consumer. So even though the service is offered for free, companies providing such services collect revenue from advertising and secondary transactions with data brokers. Since there is often no immediate monetary cost to the consumer, it is easy for the consumer to accept the good or service. At most, one need only acknowledge reading a waiver document by clicking to accept its terms. No monetary cost, plus a streamlined waiver process, make it very easy to adopt a service that will lead to sharing of personal data down the road. There certainly are costs, but they are revealed at a later point. In any case, it is important to realize that not only the consumer but also the provider of goods and services drives forward the ever-increasing number of trade-offs of personal information for convenience.

Government agencies as well as private firms benefit from convenience-focused data exchange. Journalist Glenn Greenwald has described how the National Security Agency (NSA) under Keith Alexander followed the mantra "collect it all" as they sought new ways to gather communications data possessed by Google, Facebook, and other digital communications networks offering convenient communications technologies.[4] The idea was that there was always a wider "net," so to speak, that could be cast to capture information not already available through existing secret programs like PRISM that were eventually exposed by Edward Snowden. Similarly, certain forms of "mass surveillance," such as facial recognition scanning at public sporting events,[5] are convenient to spectators in the sense that they do not require anything from the public. People need not stop, submit to search, provide ID, or do anything else: their faces are scanned and matched against a database, and if no "hits" result nothing further will happen. So large amounts of information are collected and stored for an unknown amount of time, but the public does not notice. Subtly, they surrender personal data while they are afforded the convenience of communicating—and living—without overt interference.

Of course, the irresistibility I am describing here is not absolute. If it were, then every single person would be using email and cloud computing. And everyone who used those services would do so without the benefit of encryption or firewalls or other protective devices. Irresistibility is more factual-probabilistic than absolute, or in the words of one commentator, "contingently obligatory" rather than "logically necessary."[6] And yet there is a strong tendency to go along, not to resist, not to protect oneself as new conveniences become available. In response to widespread and unwitting privacy giveaways, we sometimes hear voices urging digital privacy-protective measures on the public. As important and appropriate as these warnings are, they are very much like educating the public about *legal* rights. A person possessing knowledge of law and legal procedures can protect herself against predatory lenders or unscrupulous bureaucrats in the same way that a privacy-savvy person can minimize data privacy harms for herself. Fraud and abuse would occur less often if more people knew how to avail themselves of legal remedies, and educating the public in that regard is vital, just as privacy self-help is critical for the public to learn. Nonetheless, we cannot expect all or most members of society to attain that level of efficacy. Privacy laws—like consumer protection laws—must supplement the prudence developed by individual members of the public.

Hidden Persuasions and Soft Nudges

As consumers make choices that they believe to be based on convenience and efficiency, they are subject to influence from outside. One of the first studies to explore the influence of advertising in particular on consumer choice was Vance Packard's classic 1957 work, *The Hidden Persuaders.*[7] Packard looked closely at the collaborations between "motivational researchers" and the advertising industry. Using varied approaches that included focus groups and depth psychology, researchers helped advertisers to determine not only what products people were likely to buy but also what kind of appeals would be most efficacious in selling those products. Some of the conclusions were unexceptional: for example, it would not shock anyone to know that sad or negative images and associations are bad for sales. Much more troubling was "studying the housewife's menstrual cycle and its psychological concomitants in order to find the appeals that will be more effective in selling her certain food products."[8] Likewise, even some commentators at the time balked at the host of a preschool kids' show who hawked vitamin pills by crooning about how pretty and easy to swallow they were.[9]

Having determined that consumers behave irrationally, motivational researchers set out to use that knowledge to perfect sales appeals. As Packard puts it: "Business Week, in commenting on the often seemingly irrational behavior of consumers, said, 'People don't seem to be reasonable.' However, it made this further point: 'But people do act with purpose. Their behavior makes sense if you think about it in terms of its goals, of people's needs and their motives. That seems to be the secret of understanding or manipulating people.'"[10] Packard's study is more than six decades old now, and in some ways both consumers and advertisers have become more sophisticated. Moreover, the book has long been criticized for its lack of causal evidence—that is, what did the discoveries about consumer motivation cause sellers and buyers to do? Which ad strategies actually shaped behavior and how?[11] Because Packard's text mainly takes the form of quotes and summaries of research projects drawn from interviews with scientists and advertisers, it is short on empirical evidence in general. That is certainly a flaw in an academic work. At the same time, though, a remarkable amount of the consumer manipulation described by Packard in the late 1950s continues today, despite the advances in consumers' rights, women's rights, and the amount of information available to the general public via the Internet. Today, it's hard *not* to notice ads mobilizing fear, glorifying abuse of women, and peddling addictive substances.

Packard concludes with a plea for the protection of autonomy. He says that the biggest threat posed by the "depth manipulators" is that "they try to invade the privacy of our minds."[12] That right, which he calls "privacy to be either rational or irrational," is under threat and must be protected. Here, though he does not say so explicitly, Packard is describing autonomy. It should be up to us to make good or bad decisions about what we buy. But he leaves us with a tantalizing problem. Reading through Packard's book, we encounter one manipulative sales pitch after another. We are troubled, as he surely intended that we be, to learn of the ways in which advertising works "behind the backs" of consumers. The solution would seem to be limiting the operation of these subconscious appeals, making advertising more honest and straightforward. But Packard tells us that the "privacy of our minds," which might lead us to choose irrationally, must be defended, even if it leads us to choose irrationally (which is what depth manipulators want consumers to do). So it is a question of welfare versus autonomy. When welfare and autonomy collide, which should win?

Cass Sunstein takes up the problem of influences on choice (or in his terms, "choice architecture") and their effect on autonomy in his recent book, *Why Nudge?*[13] He approaches the problem as one of policy design: should government seek to counteract harmful irrational behaviors that arise from advertising or elsewhere? Sunstein does not confine the cases considered to surveillance and privacy, or for that matter, advertising. Rather, he utilizes examples ranging from automobile fuel economy standards to dietary options. Sunstein is arguing in favor of a "soft paternalism" in public policy. In certain cases, he supports the idea that government can "nudge" people toward consumer choices that promote their welfare, with "welfare" defined broadly as favorable health outcomes or monetary savings. "Hard" paternalism, in which people are compelled to act in a particular way and no room is allowed for choice, is always more difficult to defend as a policy orientation, but soft paternalism has its place, Sunstein says. And it is already in use, in some forms. Dietary warnings and nutrition information on food packages, as well as fuel economy facts on new car labels, nudge consumers toward better choices.

Autonomy comes into play for Sunstein as an objection to soft paternalism. He notes that the kinds of "nudges" he is advocating might draw libertarian objections based in autonomy. Don't individuals know best how to choose for themselves? Sunstein actually breaks this objection down into separate parts. For one thing, it is an objection to many nudges that we already get and yet do not protest. "Choice architecture"—"the background against which choices are made"—is an inescapable phenomenon. It "effectively makes countless decisions for us, and it influences numerous others, by pressing us in one direction or another."[14] If nudges are given constantly already, without objection, how seriously should we take the autonomy argument when it is used to attack soft paternalist nudges?

Second, Sunstein addresses the epistemic argument for autonomy: simply stated, people are in the best position to know and to choose what is best for them, because they know themselves best.[15] This is an empirical argument that may prove true for some cases and not others, he suggests. Returning to Packard for a moment, we can find ready examples of "hidden persuaders" who make us think we know what we want, even though we have less information than the persuaders do. Thus, we cannot know in advance in every case that we have enough information to make the choice that maximizes our well-being. Sometimes we do, but sometimes we do not. And finally, Sunstein suggests that some of the objections that are purportedly about autonomy are really objections about welfare. At a certain level of abstraction (e.g., saying that people want "things to go well")[16]—autonomy and welfare are hard to differentiate. If welfare (or well-being) is assumed to be the *end* that people are seeking, then a paternalism of *means* becomes less objectionable. This brief exploration of outside influence on consumer choice in the works of Packard and Sunstein provides a foundation for thinking about the bargains of convenience and efficiency that consumers face in the surveillance society.

The Allure of Convenience and Efficiency

The benefits of convenience and efficiency are widely celebrated. Advertising spots promise savings of time and money and tout the advantages of doing banking from a mobile phone and storing data on ever-smaller devices. Such messaging spurs an endless search for more advanced convenience and ever-greater efficiency. Individuals respond affirmatively to these promises out of a combination of "fear, familiarity and fun."[17] There's nothing inherently wrong with convenience, of course. And some conveniences are vital in life-and-death terms, such as laser surgery, wearable insulin pumps, or machines that constantly monitor blood pressure and heart rate in a hospital room (making it unnecessary for the attending nurses to get up repeatedly and take readings). The question, rather, is whether convenience or efficiency has eclipsed other important values, such as privacy, autonomy, dignity, or even safety. Trade-offs between these values can happen knowingly and deliberately, or they can be so poorly understood that their harmful consequences are unanticipated.

The lack of understanding of consequences is tied directly to convenience/efficiency. Explicit, written privacy policies exist to govern companies operating websites and platforms. All Internet search providers, retailers, and social media companies have some form of privacy policy, and those policies can be accessed easily. But it is inconvenient to actually read them. They are typically long, complex, legalistic, and opaque. In one study, researchers concluded that Internet users would need to spend 244 hours per year (or 40 minutes per day) to read the privacy policies for the websites they visit. A user would need eight to twelve minutes to read just one.[18]

In addition to the challenge (or inconvenience) of reading privacy policies, researchers have discovered a lack of information and a poor understanding of information collection and privacy risks on the part of the public. In a 2003 survey conducted by the Annenberg School at the University of Pennsylvania, 40 percent of respondents said they did not know how cookies work.[19] Cookies are an important device in Internet data storage and retrieval. A cookie is a bit of text that is sent to the user's computer by the web server, and the next time the user accesses that server, the server receives the cookies back, thus creating a browsing history for the user's computer. Obviously, the history can then be tracked. Not knowing about cookies is the virtual world's equivalent of not knowing about closed-circuit video cameras in physical space. In addition, 57 percent of respondents believed (incorrectly) that the fact that a website has a privacy policy means that "the website will not share customers' information with other web sites or companies." Of course, what matters is not the existence of a policy but rather what the policy actually says. Sixty-four percent said that they "almost never" look for information on how to protect online privacy. Forty percent of adults who use the Internet at home also reported that "they know 'almost nothing' about how to stop websites from collecting information about them."[20] While these findings show that many Internet users are being imprudent with regard to protecting their personal information, there is more to say than that. This system is structured in a way that makes it easy for people to compromise their information and to make bargains and trade-offs without complete information, which skews the bargaining relationship. Moreover, the data collectors benefit financially from consumers' imprudence and ignorance because it provides them with something to sell. Consumer protection legislation is one appropriate response to this situation. The European Union's General Data Protection Regulation, which took effect on May 25, 2018, requires, among other things, an affirmative step by Internet users before they access a website. For example, they must demonstrate an awareness of the cookie policies of the server. A prompt appears on the screen to which the user must respond before continuing to view the website. Other provisions include prohibitions on data sharing and a "right to be forgotten," which requires deletion of personal data pertaining to users. This step of requiring deliberate disclosure may seem small, but it pushes Internet users to make more conscious decisions about what data to share.[21] It also flips the presumption that use of the Internet and social media signifies a willingness to surrender control of personal data: data privacy is now declared by the EU to be a fundamental human right, and data remains private. Data can only be shared when the right is affirmatively waived under certain specified conditions. Interestingly, the EU law does not ask consumers to sacrifice much in the way of convenience; they must simply indicate their conscious decision to share data, and their awareness of privacy risks, by clicking on a prompt. In any event, digital privacy and private profit have lined up rather differently in US law, where no such comprehensive regulatory scheme has been enacted and where the default is loss of privacy rather than its protection.

The aforementioned Annenberg survey is a decade and a half old now, but the Pew Research Center has been following changing norms and attitudes among Internet users over time. A 2017 Pew report noted that the number of people who use social media has increased dramatically since the 2003 survey, but also found that "91 % of Americans 'agree' or 'strongly agree' that people have lost control over how personal information is collected and used by all kinds of entities."[22] Pew also collected open-ended responses to questions about trust and social media use. These qualitative data consistently support the view that users do not trust platforms to safeguard their personal information. Yet despite this gap in trust, people avow continued use—often because of convenience. "Trust takes a backseat to convenience for most," as one respondent put it.[23] Another respondent was even more blunt: "Trust is irrelevant. We know that people are wildly uncomfortable with the amount of information that, e.g., Google, has about them, but it does not stop them from using Google. People need to live their lives and they will use the services they find necessary."[24] Again and again, survey respondents freely acknowledged their concerns about information privacy on platforms but affirmed their intention to keep using them. And convenience was often the stated or implicit reason. This skeptical or ironic stance is made graphic when users wear "Google Privacy Policy" T-shirts, which are widely available for purchase on the Internet. The T-shirt phenomenon is a fascinating display of the negotiation between agency and powerlessness that (post)modern individuals engage in every day. They feel that their personal information is out of their control (powerlessness), but they choose not to change their habits (agency); they feel hopeless about changing the situation (powerlessness), but they buy T-shirts that joke about their predicament (agency).

We are really talking here about several kinds of convenience at once. The first is *convenience in the course of meeting an external demand.* When the citizen or consumer is asked to do something that is (putatively) necessary, that "necessary" intrusion is made easier and therefore—arguably—less objectionable. For instance, when security officials decide that added measures are necessary to prevent a terrorist act at a public event, they are asking something of the citizenry, even if the intrusion is for the good of the citizens themselves. So convenience here amounts to offering the public a less burdensome alternative: rather than stopping and inspecting every person, facial screening is performed as the public goes about its business. No one need change their behavior or miss a step—unless they happen to fall under suspicion. There is no need for formal consent, and the only way to visualize a bargain here is to think of the possibility of objecting. If anyone knows that facial recognition is in use, and fails to object, we can say that some part of the public has deemed the trade-off to be worthwhile. Trading safety for intrusion is acceptable when the intrusion is minimal, as with remote facial scanning. But this is a big "if," as most people will not be aware they are under scrutiny and therefore cannot be said to be consenting to it.

A second form of convenience is the *desired ease of effort regarding something that consumers already want to do.* This is the most obvious and familiar species of convenience at issue here. People want to search for information with the least amount of effort; the Internet offers them that convenience. People want to store hundreds of photographs on their phones; data storage features make it possible to do so with greater convenience. Security theorist Bruce Schneier references this type of convenience when he says, "Convenience is why we allow corporations to invade our privacy. We give companies our data because the results improve our quality of life."[25]

The third form of convenience that we see in these trade-offs might be called *second-order convenience,* because it is related to the process or transaction whereby we obtain the primary conveniences such as electronic mail and cloud computing. In other words, the thing that we want because it will "improve our quality of life" is convenient to obtain. It is simply a matter of installing an app or signing up for a service online. This second-order convenience makes the privacy trade-off even more seductive because getting it requires so little effort.

After drawing out the types of convenience named above, it is also necessary at this point to say a word about how the terms "convenience" and "efficiency" are being used in distinct ways here. Their meanings are similar but not identical, and while they are sometimes used interchangeably, they carry different emphases. *Convenience* connotes ease or lack of effort. A convenience enables one to do the same task with less effort, or to go about one's regular activity without extra effort or change in routine. In what follows, I will utilize the notion of convenience when talking about labor-saving or effort-saving innovations. Often, when people opt for convenience, they have not considered fully the ramifications of their choice, but are lured by the ease of it. *Efficiency,* in contrast, has to do with *savings*—of time, of money, of materials. Emphasizing efficiency often suggests that a person has made a calculation, even if the calculation is wrong or incomplete. Of course, it is possible to describe some actions as convenient *and* efficient, when an action results both in less effort and in a savings of some sort. But focusing on one or the other of them at a time in this section allows us to gain specific insights related to technological bargains. Next, I will examine airport prescreening programs in terms of convenience, and customer loyalty cards in terms of efficiency.

Convenience: Prescreening at the Airport

When one considers the types of convenience-related bargains arising in contemporary social life, air travel comes readily to mind. It is a complex and time-consuming process that has become even more challenging to navigate in the post-9/11 world of heightened security protocols and securitizing technologies. The complexity is concentrated at the airport, where passengers are required to pass a security search on the departing end and then clear several steps of questioning and examination on the arriving end. Since air travel is a necessity for so many people, anything that makes the process more efficient will draw interest. Various "trusted traveler," "registered traveler," "known traveler," or "preclearance" programs were implemented post-9/11, and they had one thing in common: the offer to engage in trade-offs between privacy and speed.[26] In this section I will explore this convenience-driven bargain and consider the extent to which it becomes irresistible. I will focus on registered traveler programs and compare the United Kingdom's "Registered Traveller" program with the "TSA PreCheck" offered in the United States. Taking up Peter Adey's challenge to investigate "just how these sites are experienced, not just by scholars and academics but by those who do so on an everyday occurrence,"[27] I applied for TSA PreCheck by completing an online application and submitting to an in-person interview. Below, I report the results of my application and I also relate my impressions of the experience. It involved not simply a surrender of personal information, but an emotional reaction, a constellation of feelings about social status and identity.

First, though, it is useful to consider what other scholars have said about the place and function of airports in contemporary society. The airport is sometimes described as a non-place, a liminal space through which people pass en route from one location to another.[28] But it has also been pointed out that attention should be paid to what goes on within the airport, to what the airport *does*. In other words, the airport is a place that people enter, and its specific features act upon and affect them. As a built environment, airports have been designed to evoke certain feelings and produce certain reactions. Placement of light, color, and windows are meant to produce feelings of movement toward a destination, of excitement, of a world of possibilities. As Gillian Fuller puts it, the airport offers "the promise of an actual elsewhere," an "unmediated spectacle of movement" that combines "the putative qualities of 'lightness' with utopian fantasies of flying."[29] Such design choices serve to distinguish the airport from other structures, such as city government buildings, that are experienced as static and routine-bound. Contrast, for example, the drab, hard-edged, cold, colorless, and forbidding appearance of the interior of a public welfare office with the bright and airy corridors of a high-traffic international airport. One is degrading, the other inviting. In a welfare office—or at least, in every welfare office I visited when I worked as a welfare rights lawyer—the walls were bare except for informational posters. Entrances were guarded, and seating was uncomfortable. The workers and security guards were grim. There were no conveniences or amenities. At the airport, by contrast, colorful and sophisticated advertisements urge the traveler to dream about a destination even more exotic and inviting than the one to which they are presently flying. Shops offer all manner of consumer goods.

That is not all there is to it, however. In a more functionalist sense, the deliberate and restrictive channeling of movement through the airport serves the dual goals of enhancing security and facilitating commerce. Entering and exiting, travelers are required to pass inspection points where they can be surveilled individually, and this, at least in theory, serves to secure the airport (and the planes, and the interior of the nation) against dangerous travelers. At the same time, lining the airport passageways with shops urges travelers to be consumers, to engage in commerce during their short time inside the airport. These two ends—security and commerce—are served by airport design and logistical planning. Although they sometimes stand at odds, security-related and commercially related features of the airport make up its particular version of a "surveillant assemblage."[30] The airport is not a unidimensional or conspiratorially controlled institution, but a site of convergence of multiple intentionalities and processes.

Although it is an assemblage, the airport is also a machine, or set of machines. As a "difference machine,"[31] it sorts and reassembles travelers in a variety of ways. The work that the machine does is to produce differences. Some travelers are permitted to move rapidly and unobstructed through the airport space and to fly around the world; they are "kinetic elites."[32] Other travelers are closely inspected and even, in some cases, refused permission to fly. The airport's inspection machinery performs this difference-producing work. People who pack liquids in their carry-on bags are differentiated from those who don't. More nefariously, passengers who conform to a particular physical type (read as dark-skinned, or exotic, or Arab) are differentiated from travelers seen as nonthreatening. Names that appear on "no-fly" lists are differentiated from those that don't, and their bearers are barred from flight. To put this point graphically: a stream of humans approaches and enters the airport. They arrive as motorists or bus/train/taxi passengers. There is little to distinguish them until they enter the airport space, and more particularly, the security checkpoint. There, they are sorted and differentiated along the lines indicated above. Some will pass easily; some will be questioned; some will be stripped of belongings; some will be barred from further passage onto airplanes. Though it may seem obvious to say so, this processing is the work of the airport-as-machine.

The machine not only differentiates one person from another but also fragments each person whom it encounters. Through baggage inspection and stowage, travelers are separated from their personal property when the plane loads and leaves, and reunited with it upon landing. In the words of one commentator, the airport disassembles passengers and later reassembles them.[33] This is another type of difference-production: differentiation among parts. Taken together, these two kinds of production of difference (travelers separated from each other and travelers separated from their parts) serve to underscore the point that airports are more than non-places, more than replicas of social relations determined elsewhere. Rather, they are sites of differentiation, of filtering.[34] As filters in a drinking water system produce a sanitized product, so do airports as filtering mechanisms remove certain elements from the social body. The key difference, of course, is that biological/bacteriological distinctions drive the former process while the latter involves political judgments. We have seen again and again how on-the-ground judgments as to passenger processing (and threat assessment generally) merely reinscribe biases related to race, ethnicity, and socioeconomic status. Often these biases fly in the face of facts, as when the Unabomber in the mid-1990s was widely assumed to be Arab (which he was not), or when the bombers of the Oklahoma City federal building were assumed to be Arab terrorists (which they were not).[35] Those examples demonstrate the danger of ascribing scientific validity to what are in fact political judgments, or treating social-political patterns of interaction identically to the study of biological processes such as one would be able to conduct in a laboratory. The design of the airport machine allows for subjective judgments by the human actors who guard and control its switch points, and in that way the machine falls short of the efficiency to which it aspires. When Transportation Security Administration (TSA) officials decide whom to scrutinize more closely, they are often reinscribing patterns of difference that originated outside the machine, in the social imaginary.[36] The objectively valid production of difference promised by the designers of the screening system fails in this respect. And in a broader sense, it falls short of the 100 percent fail-safe, the "sure shot" that technocratic ideology promotes.[37] The idea that a perfect airport design and screening system can guarantee and eliminate all threats is a chimerical fantasy, even if state security services do foil terror plots and neutralize specific threats in some cases. And the government acknowledges as much. As we are asked to submit to screenings and searches, the state disclaims any ability to keep us safe, calling into question the quality and advisability of the bargain in which we trade liberty for security.[38]

The dual ends of political economy and security that are served by the airport in its design and function tend to create grooves and order behavior in the same way that platforms do. Customers use Amazon to buy one product, and it becomes easier to use it again for the next purchase rather than starting the process of entering personal information and credit card details elsewhere, at another site.[39] Frank Pasquale explains the self-reinforcing effect this way: "If I want garbage bags, do I really want to go over to Target.com to re-enter all my credit card details, create a new log-in, read the small print about shipping, and hope that this retailer can negotiate a better deal with Glad? Or do I, ala Sunstein, want a *predictive shopping* purveyor that intimately knows my past purchase habits, with satisfaction just a click away?"[40] In this way consumers are habituated to returning to the same platform each time they need to buy something, and the repeated behavior wears the grooves even more deeply. Of course, the habituated behavior is expected by those who own and design the platform. And the same is true of airports as their design channels flows of people in certain directions where they can be observed and simultaneously invited to consume, to engage in commerce. Airport architecture and mobilization of space is intentional, and anyone who has passed through an airport can readily recall how shops had been placed in their path seemingly at every turn. Designers construct a path of experience in the airport via architecture.

Of course, the airport-as-machine would be less likely to exist and persist and harden as a feature of contemporary social life if it did not offer something that made people willing to tolerate it. In a democratic society, institutions that impose a burden on the public have a greater chance of surviving when they serve a purpose that is endorsed by the citizenry. We need to travel, we *want* mobility, and the differentiation produced at airports affects travelers differently. For some, the process can be made more convenient; all they need to do is submit to a front-end screening in advance, and if they pass it, their airport experience will be streamlined. This convenience/privacy trade-off arises as a means to avoid impediments to mobility, and for those who depend on flying, the trade-off often appears irresistible. This is a specific desire of individuals who do not wish to spend hours in line before boarding a plane, but it is also a tendency that reflects and strengthens a set of social relations rooted in the capitalist mode of exchange. "Mobility is a fundamental feature of the flexible capitalism that now dominates the world of exchange, production and consumption," writes David Lyon.[41] In other words, we want mobility as air travelers seeking an efficient passage from one airport to another, but we also "want" it as actors in a system where proffered services are tailored to a convenient/efficient consumer experience, and where businesses survive and flourish only when they can deliver a service that meets consumers' ever-changing specific expectations. Nick Srnicek reminds us that while the relations of production have changed dramatically over the time period of capitalism's existence, capitalism is still fundamentally about private profit. Firms emerge, compete, and diversify in order to generate greater profits than they did previously (and greater than what their competitors can attain). Pursuit of profit is the irreducible imperative of the mode of exchange and production in which we continue to live, no matter how much the social relations making up that system have changed over the past century.[42]

Here is a side-by side comparison of the UK and US registered traveler programs. They are substantially similar.

Program 1: Registered Traveller, UK Border

Cost of program: Online application costs £70; however, applicant will be refunded £50 if his/her application is denied, with yearly membership renewal costing £50. Additional costs to add children to membership: £20 for each child plus £2 a month for each child until parent's membership expires. If application is denied, membership fee is refunded.

Application process: Submit an online application form (five–ten minutes) and receive a decision within ten working days. Once the application is accepted, traveler must go through "other passports" lane during the next flight and fill out a landing card. An immigration officer will check documents and inform the traveler whether full membership has been granted.

Immediate and explicit privacy costs: Applicants must reveal passport and credit card details for the online application and submit to background checks.

Benefits: Allows traveler to get through the UK border faster by using UK and EU passport entry lanes or ePassport gates (if your passport is biometric/chipped). Also relieves traveler of the requirements to fill out a landing card, undergo an interview, and provide fingerprints (for visa holders). Registered traveler status is available to residents of UK and visa holders from certain countries It can be used at most airports in the UK and at certain Eurostar (train) terminals in the EU.

Program 2: TSA PreCheck

Cost of program: $85 for a five-year membership.

Application process: Submit an online application (about five minutes), then schedule an appointment at an enrollment center, where background check and fingerprinting (around ten minutes) will be done. Successful applicants receive a Known Traveler's Number.

Immediate and explicit privacy costs: Applicants must reveal passport and credit card details for online application and submit to background checks and fingerprinting. Eligibility is limited to US citizens and US lawful permanent residents; also, certain criminal offenses will disqualify an applicant. Children twelve years or under can use parents' TSA PreCheck without being a member themselves, but children thirteen years or older need their own memberships.

Benefits of program: Faster passage through airport security by not having to remove shoes, laptops, 3-1-1 liquids, belts, or light jackets. TSA PreCheck is only available when flying with one of the thirty-seven participating airlines to or from one of the 200 participating US airports. This program is an option for some and not others. As Benjamin Muller puts it, "For those with the means and no fear of having their backgrounds examined, the experience of the biometric border is arguably more of a possibility than a limitation."[43]

In the summer of 2018 I completed the two-step application process for TSA PreCheck. The first step was a simple online application asking for routine personal information such as name, address, and residence. The application also prompted me to indicate whether I had been convicted of a crime or hospitalized involuntarily, and whether I had been living at my present address for more than five years.[44] When the application was complete, I was directed to a processing center for the second phase, which would consist of a personal interview. I did not know how fine-grained or intimidating the interview would be, but as it turned out the process was quick and rather mundane.

I reported to the site at 9:00 a.m. and finished by 9:20, even though I had walked in without an appointment. The site location was a strip mall less than two miles from my house, where a private company offered processing of various types of identity-related documents. The questions were perfunctory and mostly duplicative of the ones I had already answered online. The clerk who conducted the interview was pleasant and friendly; we even talked about the recent heat wave and about vacation spots. Of course, I do not know what other TSA PreCheck applicants have experienced; I can only report what it was like for me. My physical appearance (as a blond-haired, middle-aged male of European ancestry) may have generated a particular response; I simply don't know. In any event, there was no searching inquiry, no request for travel history, and no open-ended questions. The clerk did ask if I traveled often outside the United States, but she asked in the context of offering a mobile app that could be used for entering certain airports outside the country. I provided my fingerprints (all ten) through an electronic device on the desk where we sat across from each other. Probably I did not find the fingerprinting to be as intrusive as it would be for someone else because I had already been fingerprinted years earlier as part of my state bar application and also as part of a background check for youth sports coaching. After the brief interview, I paid the application fee and waited for a decision. If approved, I would be provided with a "known traveler number" that could be entered on boarding passes and frequent flyer accounts; the number would flag me for expedited airport screening in the future.

A privacy statement accompanying the application informed me that by consenting to use the program I also agreed to allow my data to be submitted to the Department of Homeland Security (DHS) as part of its continual threat assessments. There was no real choice involved—if I did not agree to data-sharing with DHS I would not be able to receive TSA PreCheck. One is normally not in a position, when bargaining with government, to change or dictate terms. Nonetheless, I had to decide whether or not to consent. My ability to obtain expedited airline passenger screening was conditioned upon my acceptance of the information sharing. The statement also informed me that my data would not be shared with any private entity, although the office where the screening itself took place was itself a private entity.

About ten days after my interview, I learned that my application had been accepted and that I would be enrolled in TSA PreCheck, though my eligibility could be terminated, in the future, upon the occurrence of certain events. The next time I flew, I presented my "known traveler number" at check-in and was permitted to bypass the normal security line. As previously noted, a PreCheck passenger is not subject to the "3-1-1" rule for liquids or the requirement to remove shoes and belt. Writing this, I am struck by how minor those benefits are in themselves, though the ease of movement through the shorter line is perhaps more significant. Also, the colleague with whom I was traveling on the flight was also processed and treated as a PreCheck passenger, even though he had not gone through the application process; it seemed that he obtained the benefit as a result of traveling with me.

Nikolas Rose offers a helpful way of thinking about social relations that applies to registered traveler programs. Rose suggests that access to goods, benefits, and status are regulated through circuits of *inclusion* and *exclusion*. Both facilitate social control, but they work in different ways.

Circuits of *inclusion* operate "through conditional access to circuits of consumption and civility: constant scrutiny of the right of individual access to certain kinds of flows of consumption goods; recurrent switch points to be passed in order to access the benefits of liberty."[45] These benefits include immigration, credit, housing, benefits, consumer goods and discounts, and social status. The switch points where data is collected and subjects are evaluated sit within the state apparatus in some instances and outside that apparatus in others. Different kinds of professional experts are charged with guarding the switch points: architects, insurance companies, trainers, and management consultants, among others.[46]

Circuits of *exclusion*, by contrast, operate to place certain individuals or groups outside the circuitry by which desirable goods and social status are obtained. Designations such as "unproductive," "criminal," or "high-risk" signify unfitness to compete for goods and participate in society: they mark categorical exclusion. Both kinds of circuitry rely on data collection to regulate individuals' movement through them. And both kinds of circuits function in inconsistent and even contradictory ways. Moving through them, individuals engage "in a diversified and dispersed variety of private, corporate and quasi-corporate practices, of which working and shopping are paradigmatic." Rose continues, "These assemblages which entail the securitization of identity are not unified, but dispersed, not hierarchical but rhizomatic, not totalized but connected in a web of relays and relations."[47]

The TSA PreCheck program constitutes one of these switch points. A successful applicant gains access—that is, inclusion—to a set of benefits that have to do with mobility. They can pass more easily through airport security and receive better treatment in the process. It is easy to envision the wider "web of relays and relations" in which travelers exist when we consider what kind of person—or subject—is likely to be successful in seeking access to PreCheck. A successful applicant must:

- Be aware of the program.
- Have the money to apply.
- Have an uncomplicated residence history.
- Have an uncomplicated criminal history.

Each of these criteria depends in turn on other circuits that form the web of interconnected social life that we live in. Access to credit affects residence history; level of law enforcement scrutiny affects criminal record, and so forth.

By submitting personal information in advance, travelers gain access to a faster screening and a less invasive search process. The circuit through which air travelers pass can include or exclude would-be travelers. Exclusion awaits travelers who get placed in certain suspect categories, for example, "transient" or "out-of-the-ordinary."

Notably, the TSA encourages all travelers to apply for PreCheck. Posters inviting applicants can be seen all over airports, and the TSA site advertises the program as well. Of course, applying does not guarantee acceptance, but in theory many more people could qualify, making the PreCheck status less exclusive. Mass increase in PreCheck enrollment would change the dynamics of the airport security experience, as it could makes searches quicker for everyone by decreasing the number of travelers subject to a full search. No matter how the numbers of PreCheck/non-PreCheck travelers balance out, one thing is clear. The quantity of personal data collected by the government (or collected by private contractors and supplied to the government) will increase significantly as a result of PreCheck. Everyone who applies for PreCheck supplies data as part of their application whether they are approved or not, and those who *are* approved will thereafter leave a data trail in the course of all their air travel. And of course, much of the data has been or will be collected by other means, but the point is that a distinct set of identifying data—fingerprints, residence history, photo, arrest record, air travel history, and so on—is now aggregated and stored in one very easily accessible government database belonging to a federal agency that regulates movement. As we know, composite data has much different privacy implications as compared to discrete, atomized, and scattered data points. So the trade-off certainly incurs privacy costs. These costs, though real, are difficult to quantify, as I may never know how much data has been collected by the TSA or how it was shared.

One thing that PreCheck does is to elide some of the questions that arose concerning other, earlier passenger screening programs. The passenger screening system in place on September 11, 2001, and continuing for a time after that, focused on data gathered at the time a given flight reservation was made.[48] The data was ostensibly destroyed after the flight in question actually took place. That program was eventually discontinued, but TSA PreCheck now makes the question of record retention irrelevant. Applying for and joining PreCheck creates a record that is permanent by definition; it is the condition of continued membership.

It is a simple matter to evaluate this information about TSA PreCheck in terms of the bargaining-related questions I posed in the previous chapter:

- *Who is posing the bargain?* The government offers this bargain through the Transportation Security Agency, which falls under the Department of Homeland Security.
- *What is the balance of bargaining power?* As in most instances where we transact with the government, members of the public lack bargaining power. We can choose to apply for PreCheck or not, but we cannot pose alternative terms or modify the bargain in any way.
- *What is surrendered/gained?* We provide the government with access to our personal information in a format that makes it very easy to track our travel history. Travel history in isolation is not terribly private, but it can reveal a lot about us when combined with other data, such as communications metadata or content. In return, we can travel more conveniently through airports and experience slightly less extensive scrutiny at the airport security checkpoint.
- *What is hidden?* Our travel information will be used to conduct threat assessments; this much we are told. We do not know how often those assessments will be conducted, or how our personal data will be used to conduct them. If an assessment concludes that a traveler poses a safety risk, that passenger may be barred from flying and might not learn why.

The gains afforded by TSA PreCheck amount to a time savings and a slightly less invasive in-airport public screening. In addition to those tangible benefits, though, it is important to consider the affective, status-related differences experienced by a successful TSA PreCheck applicant. The simplicity of the application process and the speeded-up movement at the security checkpoint evoked, for me, a feeling of relief. I responded to the friendly manner in which the officials treated me, both at the interview and the airport. The experiences struck a contrast with other experiences I have had while interacting with government or security personnel. At the airport, I was also aware of how other airline passengers seemed to view *me* as I was directed toward the TSA PreCheck line with its accelerated screening. At the least, they were curious about how we PreCheck passengers were able to bypass the more burdensome screening to which they were obliged to submit. To put it simply, I was aware of the difference in intersubjective recognition that I received, and of the resultant positive feelings even though I have a full understanding, as a social scientist, of all that is involved in the privacy/efficiency trade-off—and also, of the extent to which the behavior of the airport screeners is a scripted performance. The (small) status change is insidious, and would be more powerful, as well as unnoticed, for someone who is not immersed in the process of studying these state/citizen interactions.

Efficiency: Customer Loyalty Programs

A benefit related to convenience—efficiency—is promised by purveyors of goods and services as diverse as grocery stores, sports equipment retailers, movie theaters, and fitness centers. *Loyalty cards*, or "preferred customer" programs, are offered to (or more accurately, urged on) customers, who must sign up for the program and then present the card for subsequent transactions. Like preferred travelers, preferred customers gain benefits in exchange for ceding control over personal data, but in the latter case it is for-profit businesses, rather than the government, offering the bargain. The cardholder has access, through this "circuit of inclusion," to discounts and other savings that amount to efficiency gains in the cardholder's commercial transactions with the company offering the card program. Along with efficiency, though, the bargain promises affective gains as well, a status enhancement that cannot be reduced to time or money saved. The customer feels special, even superior.

The cards are a token of membership in a club (or a circuit of inclusion, in Rose's terms) that affords its members discounts and other benefits. In exchange for the savings, members provide data to the company in two ways: initially as part of the application process and subsequently each time they use the card (here, again, resembling TSA PreCheck). The latter collection of data via card usage provides a detailed and up-to-date purchase history, which would be of use to marketers who wish to buy such data from those who collect it. Also, and more directly, the company offering the card can "sell better" to you and others with the data they have collected.[49] Purchase records show that you bought product X, and as a result of that knowledge they can offer you more of X, or related product Y. Your purchase history leads to more effective advertising as the seller knows exactly what you want, not to mention what sales pitches are most likely to work on you. From the consumer's point of view, it is more efficient, in the short run, to use these cards and avail oneself of the monetary savings they provide, and so consumers may well ignore the long-term costs associated with them, such as unwanted email solicitation. Loyalty cards and other consumer memberships make shopping more efficient in terms of time as well. It is now possible to complete one's holiday shopping entirely online, without setting foot in a shopping mall, and it is even easier for a recognized customer to do so. It is hard to complain about a technology that helps one to avoid visiting a shopping mall, but that is sort of the point. We don't complain because shopping has been made more efficient. But saving time and aggravation comes at the (eventual) cost of loss of control over personal information and vulnerability to online marketing.[50] This privacy/efficiency trade-off is one that many of us are aware of, and yet we choose to make it. There are tangible gains, such as sale prices on supermarket items that are unavailable without membership. Thus it is easy to see why people strike this bargain so often, even if it seems inadvisable when examined more closely.

As in the discussion of TSA PreCheck, posing the specific questions about bargaining is useful in the context of loyalty cards as well:

- *Who is proposing the bargain?* The private company (grocery store, fitness center, etc.) is the bargaining partner here, as compared to the state in other instances. Nonetheless, it is usually a large and impersonal commercial entity rather than an arm's-length transactional partner.
- *What is the balance of bargaining power?* The customer is not free to shape or change the terms of the loyalty program offer but can only accept or decline it wholesale. And there is some social pressure to say yes when asked to join.
- *What is surrendered/gained?* Customers gain discounts. They may earn "points" toward future purchases or receive special price reductions by presenting the card. These efficiencies are obtained in exchange for personal information such as address, email, and phone number.
- *What is hidden?* It is difficult, if not impossible, to figure out what surrender of personal data led to a future email or telephone solici-

tation, but we do know that giving away our data, somewhere along the line, leads to increased solicitation and marketing. Targeted ads show up on the webpages we visit that are unnervingly precise, offering products related to one that we just shopped for. Even if we can't back-trace these marketing appeals, we know they result from our participation in loyalty programs.

Conclusion: Beyond Convenience and Efficiency

I would suggest, though, that the tangible cost savings and other conveniences and efficiencies made possible by preferred traveler programs and loyalty cards are not the only form of benefit they offer. Loyalty and membership cards do more than offer discounts and specials in return for personal data that will be utilized for marketing purposes. They "produce a cultural system of membership, endorse trust (in a brand/store) and establish social relations"[51] This affective register is in some ways even more important than the economic calculus when we assess what is happening with consumer behavior here. Consumers are urged to sign up, motivated by the promise of savings but also by the social pressure exerted by sales clerks, for example. One must affirmatively opt out when presented with a request to fill out an application or provide a phone number. Predictably, people do sign up, and in performing that action they "establish affective systems of affiliation, lineage and sociality."[52] When other consumers obtain loyalty cards, the pressure to participate increases. It is a social pressure, the force of a cultural norm guiding individuals into a network of relations that functions "to *socialize into ways of thinking about the material gains to be had through the cultural practice of using cards.*"[53] It is not only the acts of getting and using the cards that matter, but also the way in which we collectively think about the practice of using loyalty cards. The trade-off becomes acceptable in part because so many people are making it. But all of these pressures amount to governance as they establish persistent and recurring behaviors that structure social relations in ways that affect the efficacy of individuals and limit their ability to act and protect themselves and their personal information. The surveillance processes we experience are executed by private commercial actors as well as state actors. Governance is accomplished by both, and sometimes in concert with each other.

When Pramod Nayar claims that "we live in an age of surveillance citizenship," he is referring to the blending of consumer experiences with other forms of surveillance.[54] As he puts it, "surveillance now approximates to a technology of belonging and a mode of establishing our participation in the processes of governance of the nation."[55] What, specifically, is produced by these governance practices? A certain kind of citizen with a particular subjectivity. "A full-fledged and 'good' consumer citizen," Nayar tells us, "would be a person who calculates benefits from loyalty cards, has consistent shopping patterns and displays loyalty to the store/brand."[56] This behavior, this kind of subjecthood, is achieved through loyalty card programs. People become who the sellers want them to be, and who fellow consumer citizens expect them to be. You have a card but I don't—why not? What am I missing out on? Why was another customer in line treated more generously than me? I want those rewards to be "my rewards" too. Nayar explains that being an "'exclusive' or special customer helps to establish identity."[57] People fight over whether it is better to be a Coke or Pepsi drinker, and the stakes of that trivial argument are raised when it comes to concern who they are as people. Product preference comes to occupy the same role that hobbies and avocations play in the formation and disclosure of one's identity. The difference, of course, is that people's chosen activities suggest what they find meaningful in life, while a soft drink is a product sold for profit. And product loyalty is not claimed and displayed merely by verbal statements or by purchasing patterns; wearable merchandise helps to advertise one's product preferences, and wearing it contributes to performing one's identity in the same way that political campaign buttons do.

Once the results of a bargain are seen to include affective mobilization, evaluating that bargain in cost/benefit terms becomes much more difficult. Of course, people calculate such things in a multiplicity of ways through nearly countless micro-bargains every day: these decisions range from deciding how to treat a coworker to training for a sports event to applying for a promotion. In each case, a person's course of action is shaped by a desired end that involves some sort of status payoff: a desired work assignment, a sports victory, a promotion. Some part of the sought-after result may be tangible, but the payoff is not reducible to monetary or even efficiency-related terms. In the case of loyalty cards and trusted traveler enrollment, the bearer/enrollee gets treated a certain way by the offeror, and also sees the way peers recognize and respond to that status change. The *way I wish to been seen* appears to align with the way my peers actually see me, and I want to enjoy that status. In reality, the special treatment may extend no further than the last transaction involving the loyalty card: post-transaction, there is no reason for the cashier even to speak to me, much less throw a party in my honor. In view of that realization, the mobilization of affect that impels me toward customer loyalty performances seems misguided. But of course, affect stands opposed to reason and occupies a separate domain in human experience. My point here is simply that bargaining is complicated by the affect-related results of a bargain.

Nayar's insights regarding consumer citizenship and surveillance crystallize in his discussion of reality television. Reality TV expands the portion of time when we fall under the gaze of others. We are not observed only when we engage in travel or shopping and seek special recognition from transactional partners, but all the time. When reality TV "places people under 24x7 camera surveillance, and shows them being at ease or authentic under the camera gaze, it naturalizes the process of self-representation and monitoring."[58] It's not Big Brother who is watching (though that was the name of a reality show) but an unspecified and unknown number of viewers. Surveillance is rhizomatic; everyone participates, all the time. People are accustomed to watching and being watched.

According to Nayar, "Reality TV commodifies the emotional everyday experience of the 'ordinary' . . . protagonists and makes it do the *work* of entertainment" (italics in original).[59] And people consume it. We "belong," as Nayar puts it, to a society of surveillance citizenship, and we manifest our belonging through watching (surveilling) the lives of others through reality TV shows and by exhibiting loyal customer behaviors that entail consistent shopping patterns and willing provision of consumer data. As David Lyon points out, self-surveillance is a key component of surveillance culture. People allow others to watch them through creation and sharing of captioned photos, digital stories, status updates, and the like. These disclosures, which allow others to watch us, are made freely and voluntarily.[60] In fact, they are the material from which we shape digital selves that others can see and recognize. What people do not always realize is that these disclosures "may help to naturalize and legitimate surveillance of all kinds, to encourage new modes of cooperation of the surveilled with their surveillors."[61]

One obvious example of such "cooperation" is the use of Facebook by employers and security agencies to monitor people's activities. The unintended consequences of self-surveillance and self-making serve as a reminder of the complicated relationships between agency and structure, freedom and control in our late modern world. On the one hand, the creation of personal profiles and the narration of our daily lives is a series of free actions by which we shape, become, and disclose who we are as individuals. Thus, such activities would seem to show a degree of freedom that the digital world affords. On the other hand, when people take up the invitation to engage with the technologies of Facebook and Instagram, they "normalize," in Lyon's terms, the data sharing and data collection that fuel surveillance culture. Their day-to-day exercises of freedom lead, in the longer term, to a world in which surveillance is more central, more normal, more taken-for-granted.

One more point bears mentioning in connection with convenience and efficiency and how they often come to stand in for other things, or overshadow more complex, behind-the-back affective processes. Fear stands in a vacillating position relative to convenience. Schneier claims that we submit to private corporate surveillance in exchange for convenience and we submit to government surveillance in response to fear.[62] Thus, fear and convenience might be seen as competing motivations to surrender personal data. I have shown, though, when I offered facial scanning as an example, that fear and convenience can also work together in some cases. Security technologies that do not burden, delay, or distract us are convenient *and* fear-reducing (if we know about them). Government and its contractor-surrogates benefit from offering convenience to the public at the same time that they mobilize the public's fear.

Another complexity to the fear-convenience relationship can be seen when security-related technologies are *in*convenient. All airline passengers—even TSA PreCheck passengers—experience more intrusive screening than pre-9/11 passengers did. PreCheck makes things more convenient, but only relative to the inconveniences that invariably come with air travel in the twenty-first century. And so we all submit to some amount of inconvenience in air travel out of fear for what might happen to us without them.[63] In this instance, counter to what we see elsewhere, fear trumps convenience.

Finally, one other permutation on the fear/convenience interaction can be seen in the case of text messaging. Texting is indisputably convenient: it offers the convenience of instantaneous real-time conversation when voice cannot be used (e.g., when one or both parties are in public or in a meeting). This function is useful as it makes conversations possible in settings where silence is required. As we know, though, people use texting even when it isn't necessary. It is also used when unsafe: while driving a car, most notably. There is certainly a basis for fear in connection with texting while driving that should deter people from doing it. Many car crashes have been caused by distracted drivers who texted. Even without statistics, the risk of harm resulting from a multi-ton metal hulk traveling at 70 miles per hour while its operator isn't looking should be self-evident. So texting puts fear and convenience in the opposite relation as compared to airport searches: with texting, convenience trumps fear. Why? The phenomenon of *probability neglect* is in play here. Actors fail to assess properly the probability of harm resulting from certain activities because that occurrence is less viscerally frightening to imagine. A car accident is statistically far more likely to occur but less vivid than, say, a shark or bear attack. The "availability heuristic," which leads us to imagine certain vivid fears, results in probability neglect as we ignore more probable harms that are not as vivid.[64] Schneier notes convenience trumping fear in other situations as well: the collection of our personal data. We don't focus on the harms associated with collection of personal data because those harms are remote from the moment when we surrender the data. Harm is deferred to a later time and so we don't fear it.

This chapter has explored convenience and efficiency in connection with technological bargains. I have shown that convenience and efficiency make the adoption of goods and services difficult to resist. In this final section of the chapter, I suggested that the familiar payoffs of convenience and efficiency—ease of effort and some form of savings—are only part of the bargain. Lying behind those familiar gains are affective mobilizations: the feelings associated with enhanced status and the feelings associated with fear. Even when we think we are bargaining for convenience or efficiency, these other phenomena operate to drive us toward technological bargains as well, even if we are not fully aware of them. In familiar, everyday interactions with social media, airport security, and online shopping, our actions are inflected by fear and status anxiety, making the bargains more complex.

The obscurity of our deeper motivations for striking technological bargains is not the only obscurity we face when interacting with technological artifacts in general and surveillance technologies in particular. The ways in which data flows and changes hands is also obscure, hard for us to back-trace, follow, or assess. This problem is compounded by the ubiquity of data collection and "surveillance infrastructures."[65] The ubiquity and obscurity of the technological bargains confronting us will be the subject of the next chapter.

CHAPTER THREE

Technologies of Ubiquity and Obscurity

> Some [data] segments primarily focus on minority communities with lower incomes, such as "Urban Scramble" and "Mobile Mixers," both of which include a high concentration of Latino and African-American consumers with low incomes. Other segments highlight older consumers with lower incomes. For example, "Rural Everlasting" includes single men and women over the age of 66 with "low educational attainment and low net worths," and "Thrifty Elders" includes singles in their late 60s and early 70s in "one of the lowest income clusters."
>
> —Federal Trade Commission report on data collection

Technological forms are ubiquitous, confronting us seemingly at every turn. Time and again, we must decide whether to engage or avoid them. The sheer number of such encounters makes it difficult to refuse engagement in any sustained way, since one choice follows so closely on another. This ubiquity suggests another way in which technological engagement may be irresistible.

Paired with ubiquity is obscurity, as we remain unaware of how the ever-present technological artifacts surrounding us actually operate. In particular, we do not notice the sites at which personal information is extracted from us, nor do we see how those extractions cause us to lose control over the information. Sometimes we know that information has been shared—as, for example, when we get an email solicitation related to a product that we bought, or a request for donation after disclosing an email address online. In those instances we know that our data was shared, but it can often be difficult to back-trace how and when that happened. As a result, we navigate a society where data collection is occurring seemingly everywhere, and we remain unaware of how it happens and what its effects might be.

Ubiquitous Technologies: "Platform Capitalism"

Widespread use of readily available and easy-to-engage technologies generates a self-reinforcing effect: as people become accustomed to such electronic activities as online shopping and social networking, the demand for more of those services increases, and it becomes more difficult to find settings where they are not in use. Moreover, with the ubiquitous services come a proliferation of sites through which we can be watched and mined for data. David Murakami Wood terms this pervasive and taken-for-granted watching "ambient surveillance."[1] Where the technologies are in place, they will likely be easier to use than their pre-electronic counterparts and therefore embraced, as seen in the preceding chapter. Think of EZ Pass compared to coin-collecting toll baskets, or Amazon bookselling as compared to phoning or visiting a bookstore.

Technologies that become ubiquitous and structure our life experience have been with us for some time. Here is a personal example that illustrates their subtle force. When I lived in the southwestern United States, I did a fair amount of solo wilderness hiking. I did not own a cell phone back then, so I developed a safety-related routine of stopping in town after each hike and finding a phone booth to call home. That way, my wife would know that I had completed the trek and left the mountains. If I did not call at the expected time, then she could assume that something had gone wrong, and she would know where I was in case I needed help. Over the years, it became harder to find phone booths, and on a few occasions I did not call until hours later than planned, or even the next day. Needless to say, those lapses in communication caused her some concern as she imagined me lying at the foot of a cliff, lightning-struck, or snake-bit miles from medical aid. Eventually, I bought a cell phone. The lesson here was that it became too difficult to avoid using the communications technology that had become prevalent; the ubiquity of cell phones and related infrastructure crowded out older technologies and made cell phone use, in a practical sense, irresistible.[2] Jathan Sadowski and Frank Pasquale underscore this point about technologies that are irresistible because they are ubiquitous. And they continue: "We 'consent' by default because the options to not do things that pull us into the logics of these systems—such as not using digital platforms, not using smartphones, not going to stores and streets without a mask, not living in a populated area—can hardly be considered real choices for the vast majority of citizens."[3]

The proliferation of digital platforms in particular is noteworthy in connection with the ubiquity of technological forms. Platforms are a part of daily life for many of us, regardless of whether we know what they are or how they operate. One need not be able to define or describe a "platform" in order to use Amazon or Uber services. In fact, their ubiquity makes them hard to avoid when Amazon sells everything from books to medical equipment and Uber prices its services below taxicabs so that its drivers are seen everywhere. In line with the formulation just cited, we are hard-pressed to avoid doing "things that pull us into the logic of these systems." We confront a digital platform at so many points during the day in connection with a range of important activities such as obtaining food, moving from place to place, planning trips, and arranging business logistics. Moreover, sometimes it is the *same* platform that we encounter in the search for different necessities (or luxuries) as the logic of platform management compels their managers to grow and diversify each platform. Platforms will flourish as profit-making concerns if they grow bigger and infiltrate new markets. Thus, each of them continually offers new products and services, and we find ourselves going back to the same platform time and again.

But what, exactly, *are* platforms? Nick Srnicek offers an extremely useful definition that helps to give us a conceptual hold on this phenomenon. He writes, "At the most general level, platforms are digital infrastructures that enable two or more groups to interact. They therefore position themselves as intermediaries that bring together different users: customers, advertisers, service providers, producers, suppliers and even physical objects."[4] There are different types of platforms—five different types, in fact, according to Srnicek.[5] *Advertising platforms* (such as Google) bring advertisers and users together as users provide data and then advertisers buy access to the platform to sell to those users in a targeted manner. *Cloud platforms* rent out software to businesses. *Industrial platforms* provide logistical support to industry by managing efficiently the aspects of the production and distribution process. *Product platforms* collect fees for use of a service. Finally, *lean platforms* (Uber, for example) own as little of the means of production of the offered service as possible in order to maximize profit. In delineating and explaining these five types of platforms, Srnicek also notes that the types are not exclusive of one another—and he points out that Amazon "spans nearly all of the above categories."[6]

Platforms subsist on data. Quoting Srnicek again, "data [are] the raw material that must be extracted, and the *activities* of users [are] the natural source of this raw material."[7] This formulation emphasizes that we live now in the midst of economic conditions in which data extraction is highly profitable and conducive to the growth of firms. There are of course other ways for firms to generate profits, but data extraction is fruitful, not least because the firms themselves can increase the supply through manipulating users to yield more data. And platforms are the best mechanism to ensure profit from data extraction in a sustained way. Srnicek shows convincingly that platforms arose out of earlier economic crises and developments and are now a predominant feature of the economic landscape. Srnicek's typology of platforms bears out their predominance. Manufacturing firms rely on platforms to facilitate logistics of sale and delivery, social media providers rely on platforms to connect with other users, and advertisers rely on platforms to tell them exactly what products consumers want.

Platforms offer new services and links to other economic actors. We have seen that they tend to privatize and control infrastructure.[8] This growth mirrors consolidation and expansion at other points in the history of capitalism, from the rise of industrial manufacturing to centralization of management. As it faces out toward the consumer, economic centralization breeds ubiquity, and that relationship is not unique to the platform phenomenon. Automobiles, once they were produced on a large scale and priced affordably, became ubiquitous to the point where it was infeasible—or at least very difficult, in many areas of the United States—to function without one. Of course the infrastructural development of roads, bridges, motels, and residential communities contributed to this ubiquity and made it even harder to resist. In both cases (autos and platforms), consumers' repeated behavior tracks into grooves, as Pasquale describes it, so that the alternative paths become less desirable and the grooved paths become the norm.[9]

Pasquale suggests that the emergence of platforms ushers in a move from "territorial" to "functional" sovereignty.[10] Territorial sovereignty is a more familiar form of governance and refers to the control of the physical terrain on which the governed live and interact. Before the digital age, territorial control was the principal way to envision governance (though of course ideological battles over discourse and ideas were waged, to be sure). Platforms, however, allow their owners to attain functional sovereignty, so that the owners can shape and regulate, permit and prohibit what subjects (that is, those subject to their authority) can do. Julie Cohen emphasizes this point by using the term *transnational sovereignty* to highlight two features of the platform economy: both a user population in the billions and insulation from government regulation.[11] By whatever name, this move to a new kind of sovereignty is troubling because it is so often a private actor (Amazon, in Pasquale's example) attaining functional control via platforms. Part of the concern here is with under-regulation of platforms, which has allowed Amazon to infiltrate city government and manage so many aspects of consumer activity. In addition, the platforms are spaces where buyer and seller interact but are also owned by one of the parties to a transaction. They are "market makers,"[12] in Pasquale's terms, so that they regulate the very activity in which they are co-participants. They are "able to exert regulatory control over the terms on which others can sell goods and services,"[13] guaranteeing that the bargain will not be struck at arm's length and will be more favorable to the platform owner. Any protocol for dispute resolution, any rule for platform use, is set by the owner. This arrangement transgresses an ancient rule of politics: that "no one should be judge in their own case." This rule was set down by Aristotle in his *Politics* and cited much later by Sir Edward Coke in seventeenth-century England, and yet again by the framers of the US Constitution.[14] The idea that a judge judges badly when a conflict of interest arises is so commonsensical as to need no demonstration, but it claims a lineage of thousands of years in political theory, and the same concerns voiced by Aristotle, Coke, and Jefferson regarding self-interested judges arise with regard to market-controlling platforms today.

Julie Cohen notes that platforms flourish and grow according to logics of "intermediation" and "legibility," which I take to be, respectively, the ability to connect transactional partners and the ability of those parties to understand each other—not necessarily in terms of communication but effectively to "read," and obtain useful information about the other party. Since this is how so much transactional behavior occurs (that is, mediated by platforms), we become increasingly accustomed to interacting this way and ill-equipped to seek alternatives.

Finally, there is a wider sense, a physical-spatial component, of the ubiquity that I am describing, and it serves as a counterpart to the "virtual" ubiquity of platforms.[15] It is built into the physical environment and functions hand-in-hand with platform capitalism to limit consumer choices and decrease consumers' transactional leverage. So much has been written about the loss of public space that I scarcely need to add to it here, except for one observation. The configuration of the global city affords predictable options for engaging in commerce—so predictable, in fact, that the very same establishments can be found wherever one travels. McDonald's and Starbucks have infiltrated Paris, Kiev, Copenhagen, and everywhere else. Shopping malls are ubiquitous, popping up in train stations and city centers across the globe. These features of the cityscape enable travelers to find the same options easily in each new city. The products themselves are offered quickly and cheaply, and some people, at least, want such convenience. Cumulatively, though, these businesses make up a city where the vast majority of the physical space is occupied by commercial establishments. Access to these spaces is regulated: it requires money, and perhaps the appearance of one who is not destitute but ready to spend. Some circuits of inclusion (to use Rose's term), such as upscale retail malls and hotels, are regulated more closely than fast-food outlets and coffee shops, to be sure (although two African American men were arrested at Starbucks in 2018 after management ordered them to leave because they had not yet ordered. The men claimed they had been racially profiled—the pending criminal charges were later dropped and the Starbucks CEO apologized publicly).[16] But there is literally *no place to be* if one is not buying something. To put this point another way, one is compelled to engage in bargaining by virtue of being physically present in the city. Not to engage by shopping or buying food makes one into an obstruction, even a suspect presence.

The classical liberal notion of the freely bargaining, rationally choosing agent doesn't quite work to describe a situation where bargaining is compulsory, where people can choose among oft-replicated options, but must choose something—they cannot refuse to bargain in the first place. One thinks here, again, of the Coke "Freestyle" soft-drink machine mentioned in chapter 2, whose slogan is "Choose freely," but choice in that message means nothing more than selecting one of the Coke products offered by the machine. If the modern individual is one who freely bargains for chosen ends, what does it mean when a person must bargain—for *something*—instead of merely sitting or standing there? One cannot disengage because there is literally no place to do so. Here it seems we have become prisoners of what we demand, although certainly not everyone demands and receives this benefit of ubiquitous and fungible shopping options to the same extent. McDonald's is affordable to most; stores like Sharper Image are not. Invitations to maximize shopping benefits are not extended to all. As David Lyon puts it, "certain well- heeled groups are targeted for special deals and privileges; others, with poorer postcodes, are passed by."[17]

Once again, posing the questions raised in chapter 1 helps to crystallize the situated bargaining we experience vis-à-vis platforms:

- *Who is posing the bargain?* The platform owner connects the "two or more groups," in Srnicek's terms, whose interaction constitutes a platform. Platforms, like other means of production, are owned privately in a capitalistic mode of economic relations. Though we are bargaining with a private entity here, sometimes we are invited in indirect ways. If I already belong to Facebook or LinkedIn or Instagram, I will receive explicit invitations to connect with a new user or try a new app. That contact seems explicit, even though it comes from other users rather than from company spokespeople themselves. But the bargain is offered just as frequently through the *social environment* where a particular platform—and platforms in general—are ubiquitous. I might create a Facebook account in the first place because everyone else has one: friends, family, and colleagues. An acquaintance once told me that it was impossible, in his view, to rise in a professional career without using LinkedIn. I don't agree with that statement as an empirical matter, but I do acknowledge the normative force of such assertions in urging people toward certain techniques of career advancement that rely on platforms. In view of repeated exhortations to use ubiquitous platform technology, the pressure toward engagement may seem irresistible.[18] It is not a matter of asking whether a new social media account is worth the risk of privacy loss given its benefits. Rather, we are hard pressed to avoid doing things that "pull us into the logic of these systems," as Sadowski and Pasquale put it.[19] While the bargain we ultimately make (or decline) is with the platform owner, the invitation itself comes, in many cases, through the pressure of social norms. Millions or billions of other users recruit us to bargain with the platform owner simply by using the platform themselves.

- *What is the relative bargaining power of the parties?* Of course, we are not free to change or dictate terms when interacting through a platform—at least with regard to the owner. In the instance where a user accesses a site such as Airbnb, the terms of a rental can be negotiated, to some degree, between the lodging owner and the would-be renter. Regarding the site owner, though, things are different. Users cannot change the basic structure within which they are transacting. This feature of platform capitalism highlights the insight of Cohen and Pasquale regarding the status of platform owners as "market makers" as well as market participants. Platform owners set the rules and control the field. They are in a much stronger position than the other actors since they act as judge as well as contractual partner. This dual role also renders the bargaining situation more complex. Even though the platform is a sort of intermediary that brings transacting parties together, the platform owner benefits from the exchange—or they would not be in the platform business in the first place. Transactions generate more data as well as user fees in some instances.
- *What are the perceived/actual gains?* When a person is seeking a ride through Uber or a vacation rental through Airbnb, the actual gain is fairly obvious: the price and the purchase are plain to see. Those benefits help to support the popularity of platforms. As noted earlier, even when people do not know the economics or operational details of platforms, they use them regularly and routinely.
- *What information is hidden?* In contrast to the straightforward benefits of the bargain that come to us via platform transactions, the ways in which our personal data are collected, packaged, and sold are often

obscure. The ubiquity of platforms exists in a world where so much of the sharing of personal data is unknown or poorly understood or both.

The obscurity with which data flows operate compounds the problem of the ever-present ubiquity of opportunities to surrender data, and the volume/scale on which this exchange happens makes it all the more disturbing. The next section of this chapter addresses the obscurity of the data collection process.

Obscurity: The Collection, Packaging, and Selling of Consumer Data

There are two senses in which obscurity develops in data collection: regarding *data flows* and regarding *harms*. Data flows become obscure when people (especially in the United States) lose control over their personal information via the collection of that information and its use by data brokers. Data brokers buy consumer information that has been collected from a supplier. The data broker amasses raw information, such as individual consumers' shopping histories, and then packages that information for sale to a party interested in marketing its own products effectively. A recent Federal Trade Commission (FTC) report on data brokers summarized their activities as follows:

> Data brokers collect data from commercial, government, and other publicly available sources. Data collected could include bankruptcy information, voting registration, consumer purchase data, web browsing activities, warranty registrations, and other details of consumers' everyday interactions. Data brokers do not obtain this data directly from consumers, and consumers are thus largely unaware that data brokers are collecting and using this information. While each data broker source may provide only a few data elements about a consumer's activities, data brokers can put all of these data elements together to form a more detailed composite of the consumer's life.[20]

The sheer volume of data collected and then packaged for sale by data brokers is staggering. One of the nine companies the FTC surveyed for its report had collected information on 1.4 billion consumer transactions, and another company had amassed 3,000 data segments for every US consumer.[21] When one considers the technologies available for gathering personal and transactional data, together with the manifold interactions each of us experiences daily with retailers, courts, government agencies, social media, and others, it becomes apparent how such massive-scale data collection has become not only possible, but common. And as Torin Monahan notes, these large-scale data collection and storage practices actually make consumers' personal data vulnerable to hackers.[22]

So consumers divulge personal information as they navigate the life-world every day. They use a "member's card" at the grocery, the gym, and the pharmacy. They buy products online; they use search engines and cell phones and online dating sites. But there are consequences to each disclosure of personal data. The FTC report details the practices of data brokers and shows how consumers can suffer "downstream" harms of which they are unaware. For instance, it is often impossible to back-trace how the data broker obtained a particular piece of data since it is amassed from a wide range of sources.[23] And it is futile for the data broker to delete information from its records once that information has already been transmitted to a purchaser—deletion after-the-fact cannot recapture the data that has passed to the hands of purchasers who now own the data and can do with it what they wish.[24]

Some data is available from public sites, for example, the details of real estate purchases. Since the consumer did not provide the data to the site that makes it public, there is no meaningful consent involved. In fact, the consumer might not know that such data is available to the general public. And, of course, there are virtually no limitations on what third parties can do with the publicly available information. Other times, data is collected in consumer transactions and then provided to a third party who packages it and sells it to marketers. Such information transfer is likely to be lawful—that is, it usually does not violate privacy laws. The sharing of consumer information is vastly under-regulated. The most significant limitation on sharing it comes from the voluntarily created privacy statements of the data-gathering companies themselves. The privacy policy statement, however legalistic and opaque, details how a site user's information will and will not be collected and used. And in theory at least, the user's engagement with the platform or other site was conditioned on her understanding that those limits would be observed. To effectuate this understanding, though, the user must:

- Read and understand the privacy policy.
- Be aware of how and when it is being violated.
- Be able and willing to seek redress in court.
- Be able to prove with evidence that a violation has occurred.

This attenuated causal chain presents daunting challenges to a would-be litigant. California's Online Privacy Protection Act of 2003 is one of the few enacted laws that regulates online data collection and sharing.[25] It is primarily a vehicle for compelling site owners to display their privacy policies prominently—an improvement, to be sure, but far from an adequate solution to the obscurity problem. There are also some privacy laws unrelated to the consumer-to-company transaction that impose limits in specific contexts: the Health Insurance Portability and Accountability Act of 1996 (HIPAA) is a federal law that safeguards health information from unauthorized disclosure. But HIPAA generally applies to health-care providers, and so it does not protect health information in the hands of others.[26]

The online privacy bill of rights, which was created during the Obama administration, also states principles of information privacy protection, such as the importance of context in information sharing.[27] Data transfer must take into account the original context in which the data was collected. Additionally, consumers must consent to the sharing of their personal data before site owners may share it. These and other provisions of the privacy bill of rights are laudable and important, but they have not yet become law. The Obama administration proposed the Privacy Bill of Rights in 2012 and three years later published a report on efforts to implement it. Since then, online privacy legislation has been introduced in the Senate but has not moved forward.[28] The Privacy Bill of Rights is not self-enforcing and is not yet law. In other words, the privacy guarantees enumerated within it do not trigger enforcement or link aggrieved parties to available remedies. Thus, they merely state the aspiration to a legal protection that does not in fact exist at present.

This under-regulated legal regime, then, overlays the environment in which our online consumer activity takes place—and in which our surrendered data moves and circulates from consumer to data collector to data broker to marketer.

The FTC report illustrates how aggregated data that is compiled, packaged, and sold by data brokers can inflict harms on consumers even if the information is used legally and truthfully. For instance, the data point indicating that I live in New Jersey is innocuous by itself, as is the data point of my age group (50+). But even those two data points alone place me in risk categories for disease that could influence my ability to obtain certain kinds of insurance. This "knowledge discovery in databases"[29] makes data collection more than the sum of its parts: the assembly of insignificant data points is more valuable to data buyers and more dangerous to us that the discrete points would be by themselves. When food purchasing patterns, commuting path, and familial status are aggregated and sorted, a rather detailed "data double"[30] is available to anyone who wants to evaluate me for purposes of health insurance, government service, fitness to adopt children—and other, perhaps more nefarious purposes.

The packaging of data is something of a specialized enterprise. Brokers can offer a more attractive product if they process and package their information in highly detailed and specific ways. They create synecdochal demographic types that marketers can use to treat target markets in crude stereotype. "Urban Scramble," for example, is a designation used to refer to debt-burdened city residents living paycheck-to-paycheck. Racial and ethnic identity can be suggested implicitly or explicitly here.[31] These categories are reified as consumers are told, through advertising, that they conform to a given type. "Rural Everlasting" is a demographic label that refers to older white adults living (obviously) in rural parts of the United States who lack a formal education.[32] Data brokers create these types that confront consumers and, in a very real way, help to construct the world that consumers encounter. Frank Pasquale calls these "whole new kinds of people."[33] The creation of "whole new kinds of people" occurs in a twofold way: (1) through the composite types created by data brokers to sell their product, and (2) through the behaviors that individuals are encouraged to adopt because those behaviors are conducive to control or manipulation. A passive subject or an aggressive, appetite-driven subject can be a successful target depending on what is being sold.

Our identities are altered through the information processing done by data brokers. When data brokers create a demographic profile like "Urban Scramble" while seeking to sell marketing data, data users then come across people who have been tagged with that label and view them in a particular way: as desperate, insolvent, possibly untrustworthy city dwellers (not to mention as racialized "others"). This process engenders misrecognition by the marketers themselves, as they buy data and use it to contact actually existing people for sales purposes. Marketer A buys a data package from data broker B in order to sell products to a target demographic. The individual customer is interpellated, to use Althusserian terminology, as an example of a constructed group (Urban Scramble), and if the customer responds by purchasing the offered product, that constructed identity (or subject position) is adopted. No longer recognized as an individual subject acting for itself, the consumer is now a representative of Urban Scramble. Each time the "hailing" or interpellation occurs in the form of a sales pitch targeted at the Urban Scramble demographic, subjects become subjects by responding to the call that names them as a demographic example.[34] The more precisely accurate and successful the targeted marketing becomes, the more sharply drawn are the contours of this subject position. And the individuals named this way perform their identity again and again. If the individual rational actor is the fundamental unit of a bargaining relationship, it is inescapably problematic to note that the individual not only lacks agency but also lacks a distinct boundary of the self separating it from its data double.

In addition to the risk that personal information can be used to deny benefits to a consumer, as I have just explained, there is the potential for uses that are outright and explicitly illegal: "Identity thieves and other unscrupulous actors may be attracted to detailed consumer profiles maintained by data brokers that do not dispose of obsolete data, as this data could give them a clear picture of consumers' habits over time, thereby enabling them to predict passwords, answers to challenge questions, or other authentication credentials."[35] Compromised data is another foreseeable effect of data brokerage activities—unintended, perhaps, but nonetheless very real and dangerous. In the summer of 2017, credit reporting company Equifax acknowledged a three-month-long security breach that compromised the personal information of 143 million individuals.[36] It is impossible to know all of the effects on consumers of such a massive unauthorized disclosure, which included credit card numbers. Daniel Solove and Danielle Citron have pointed out, though, that there are some easily foreseeable effects of the breach. For one, consumers may refrain from seeking home loans and other consumer goods that entail a credit check, since they know that the data theft compromised the data about them that Equifax held.[37] Even before the breach, Equifax had reportedly been ordered by the federal government to pay millions in fines and restitution stemming from findings of unfair and illegal practices.[38] The risk that data brokers will harm consumers by intended and unintended practices is quite real and has in fact already materialized. Of course, there have been other such breaches; this one was recent, large, and well-publicized.

In addition to the obscurity of data flows, there is obscurity regarding *harm*. The obscurity of the processes by which personal data escapes our control and flows into other hands is only one aspect of the workings of the surveillance society that is difficult to grasp. Obscurity is also evident in the conceptualization of *privacy harms* associated with data collection and sharing. We don't see all the points at which data is extracted or where it goes, but we also lack a conceptual grasp of what is harmful about the extraction and sharing itself. "We" as a society fail to grasp the harm adequately, and the law follows suit. Or perhaps we could say that law *contributes* to this confusion at the same time it reflects confusion in the larger society. Legal remedies are created in response to recurring and recognized harms. Tort law, for example, exists to address the wrongful and harmful acts of others that affect us by damaging our bodies or our property. If I lose a leg as the result of someone else's careless act, my loss (harm) is easy to see and also easy to redress by expressing the loss in monetary terms. The loss of privacy that results from stolen personal data is harder to identify, much less quantify.

Solove and Citron take up this problem in a recent law review article.[39] They note the reluctance of courts to recognize data breach harms. Such harms result when data falls into the hands of someone other than the person or entity entrusted with it in the first place. This can be the result of hacking or illegitimate data sharing/collection. One important qualification must be stated here. The question of data breach harm only arises—at least in legal terms—when the sharing itself is legally wrong. It does not apply when a company sells information it was entitled to have. Thus, Solove and Citron's analysis applies only when data has been obtained illegally—there must be a legal wrong in order for a victim to obtain relief in court. So many of the instances described in this chapter where people lose control of their personal information (and suffer for it) fall *outside* of the realm of legal remedies. Data from public sites, and data that is shared in compliance with a site owner's privacy policy, do not constitute data breach and therefore do not count in the analysis of data breach harms. So data breach harms that raise the question of legal recourse are only those that result from illegal data transfer.

Within this limited universe of hacked or stolen data, though, there are still many instances of harm, and the law has been slow to acknowledge them. Because those harms do not resemble more familiar injuries that bring plaintiffs into court seeking damages, courts tend to deny relief to plaintiffs in data breach harm cases. One frequent response by courts in these cases is to dismiss a case for lack of *standing*. Standing is a legal term referring to the ability of a party to come before the court. To claim standing, it is necessary to show that one has suffered (or is about to suffer) an *injury in fact*. As the authors point out, the harms associated with data breach involve such things as fear of future consequences, anxiety about our privacy, and awareness of a heightened risk that someone might one day steal from us. What kind of injuries are these? Can we recover damages for things that have not happened yet and may never happen? Some courts have responded to these questions by simply dismissing plaintiffs' lawsuits based on a lack of standing.

Future injury via identity theft is a very realistic possibility when one's information privacy has been compromised. And increased anxiety about that or other potential bad outcomes is a form of emotional harm;[40] anxiety can be a clinical phenomenon requiring treatment. Of course, Solove and Citron are offering a way to conceptualize data breach harms that does provide remedies in the form of damage awards. They survey the past court rulings denying relief to show where the law is now and how it needs to change in recognition of the harms ushered in by the data collection industry. At times, they draw analogies to other remedy problems in law (suggesting that we are not on entirely new ground here), and at other times they acknowledge ways in which data privacy may be different. They grapple with the problem of obscurity as they look to move the law in a new direction.

One particular observation by the authors underscores the uniqueness of data privacy. After control over a piece of personal information is lost, harm can be compounded each time a data thief sells to another person. The consumer does not know when or how many times this process occurs, and in fact it is quite possible that no one knows. Thus, the magnitude of the harm may be obscure. But the *insecurity of not knowing* is itself a kind of harm. Moreover, when a consumer learns that a breach such as the Equifax breach has happened, it is quite reasonable to assume that the hackers will misuse (overshare) the compromised data. As the authors quite reasonably ask, why else would hackers steal data in the first place?[41]

Additionally, Solove and Citron note that Edward Snowden's revelations about the information shared by Google and others with the US intelligence agencies helped to concretize in the public mind harms caused by data breach and illegitimate data collection.[42] We now know that much of our routine online activity was scrutinized by the government without any particularized suspicion or reason. At least some of us will find that fact troubling, and it changes the way we think, in an ontological sense, about our relationship to government and about what, if any, spaces of privacy in our lives remain inviolate from government intrusion.[43] This is not to say that a remedy is easily imaginable, much less actionable, but it does highlight a rather vivid sense of what is being lost in this environment of constant data transfer.

As noted earlier, data breach harms are only *legally* relevant when a wrong has been done—that is, when a bad actor has hacked or otherwise stolen data. Without an identified wrongdoer, the law is powerless to intervene. However, it remains possible nonetheless to contemplate the kinds of harms (anxiety, heightened risk, etc.) that are common to lawful *and* unlawful data transfer. Some data-sharing practices may be lawful at present for online sellers, but the harms they produce are no less real. As Cohen puts it, "Networked environments configured to optimize data harvesting and surplus extraction operate—and are systematically designed to operate—in ways that preclude even the most perceptive and reasonable consumer from evaluating the goods or services on offer. Predictive profiling seeks to minimize the need to persuade by targeting directly those potential customers most strongly predisposed to buy and appealing to everything that is known about those customers' habits and predilections."[44] The harm resulting here—from lawful and routine data-extraction processes—is a harm of obfuscation, as consumers are unable to understand the bargain they are enticed to make. It is a process similar to defrauding someone, but not exactly in the way we are accustomed to thinking about fraud. In fact, it is worse than fraud in a way. Fraud results when one party to a transaction withholds important facts from the other. If the fraud is discovered, it can void a contract or give rise to remedies under consumer protection laws. What Cohen is describing is an ever-more-accurate process of learning more about a customer than the customer knows about herself. It involves predictive profiling that disables the customer from refusing or resisting the marketing offer. And it is completely lawful. Knowing that this process happens is itself generative of anxiety as we contemplate the operation of activities we do not perceive or understand.

Cohen also points out that the algorithms fueling predictive profiling are closely guarded secrets. Facebook and other companies are far more willing to share the data they collect than the algorithms by which they analyze and pinpoint consumer demand.[45] To make matters worse, the search for ever-more-accurate predictive formulae is described as science, as a search for truth, rather than as a process of improving targeted marketing.[46]

Effective marketing also exploits affect. Appeals to certain consumers identify intense emotional responses and also generate those reactions.[47] As Cohen explains, "Algorithmic mediation of information flows intended to target controversial material to receptive audiences intensifies such feelings, reinforcing existing biases, inculcating resistance to facts that contradict preferred narratives, and encouraging demonization and abuse."[48] This phenomenon is a not-quite-unintended consequence of targeted marketing. Marketers want to know what works, and if what works is emotion-laden ad content that hardens consumers' worldviews and affirms irrational prejudices, so be it. Customers who react strongly in emotional terms to a certain marketing overture might be more likely to buy the proffered product, and the continued appeals may produce more intensely emotional reactions. Market researchers know this and use it in the service of more precise targeting. Consumers are unaware of how their responses are noted and fed back into the targeting process, whether through emotional appeals to prejudice or more subtle behavioral shaping. Pasquale observes: "Once a critical mass of flags like 'I don't want to see this' or 'This is spam' amasses around one person's account, he may well be deemed 'creepy' or 'depressing,' but he may never know that, or know why the determination was made. Data scientists create these new human kinds even while altering them."[49]

The obscurity noted here by several commentators helps us at least begin to see what was objectionable to many about the infamous Facebook "emotion experiment." The company monitored, for a period of time, the responses of users to changes in the news feed on their Facebook accounts.[50] Through this experiment, the company's data scientists learned how changes in news content affected the moods (and to some extent, the behavior) of Facebook users. Company spokespeople were quick to point out that no identifiable personal information was obtained, and further, that all news content is by necessity filtered as it is impossible to stream all possible news stories at one time. So why did people object?

One objection recalls Cohen's earlier observation to the effect that the emotion experiment was not science in the form of a search for truth but rather market research by an entity with a direct stake in finding useful answers and applying them to make more money. The entity paying for the research tainted its purity. But this is nothing new. After all, the CIA began funding psychological research to be applied to intelligence gathering more than a half century ago, at the start of the Cold War.

Additionally, we become uncomfortable knowing that this data science creates "whole new kinds of people," in Pasquale's terms. The subjectivity of people living in the digital age is being altered by those who control their access to news and other information. Tech companies know something about us (in the predictive sense of how we are likely to react to emotional stimulate) that we do not even know about ourselves. This is true even though the subjects in the experiment were not individually identified. They know something about us collectively, and that knowledge can be applied in addressing each of us individually.

Restating the framing questions I've raised in each chapter once again, in connection with data brokerage, helps to illustrate the irresistibility of data sharing bargains and the consequences flowing from such bargains.

- *Who is posing the bargain?* In a sense, *no one* is proposing a bargain when data is collected from consumers. As noted earlier, when data miners look to public sites to collect information about my house or the places I've lived, they aren't bargaining with anyone. Further, the transactions whereby data brokers buy and sell data collected by site owners or other vendors are opaque to the consumers whose personal information is being traded. The only choice on the part of the consumer is the decision to enter the world of electronic communication in the first place. And even for those who don't transact online, there are other means to extract data from them, such as grocery store loyalty cards. Perhaps more accurate than saying there is a bad bargain is to say there is no bargain at all in this scenario.
- *What is the relative bargaining power of the parties?* This question, again, is tantalizing because the bargain itself is so vague. When our data is collected from public sites or lawfully sold by site owners, we lack the power to prevent such occurrences. We cannot alter the privacy

policies of Google and others, nor can we enjoin the data transfer that is constantly occurring. In view of those limitations, our bargaining power is so weak as to be nonexistent, and our only recourse is to stay away entirely from a wide range of everyday transactions both on- and off-line (or, at least, to take some protective/avoidant measures). In the case of stolen data, Solove and Citron's work on reshaping tort law to encompass data breach suggests a measure of efficacy on the part of data breach complainants, as the authors' proposed remedies find their way into court decisions. In addition, the proposed privacy legislation requiring consent before releasing consumer data would restore a measure of agency to consumers transacting online. They could calculate, at least in theory, whether consent to release information was warranted in a given situation.

- *What are the anticipated/actual benefits?* The question of anticipated benefits here refers us back to the reasons why individuals engage in e-commerce in the first place. Promises of convenience and efficiency as explored in chapter 2, coupled with the ubiquity of e-commerce as described in this chapter, lure consumers into an environment where their personal data can be extracted at innumerable points. One might say that that the benefits are recognizable but the harms against which those benefits might be compared are poorly understood. As a result, the evaluation of the bargain that consumers might do is incomplete. Moreover, even when individuals decide to provide data to market researchers, they are also volunteering information on those who share their characteristics and experiences. However, no permission and no direct benefits flow to the mass of persons the sponsoring agency learns about. There are parallels to DNA analysis here: an individual who voluntarily offers his or her information for analysis also simultaneously offers information on family members who have not agreed to this.[51] These costs are external to the individual who deems it a worthwhile bargain to shop online, answer surveys, or participate in focus groups. Nonetheless, they are consequences flowing from a social bargain. They impact others who never had the opportunity to consent or refuse, and who derive no benefits from the bargain.
- *What information is hidden?* The second half of this chapter detailed the various forms of obscurity confronting the online consumer. The consumer may be unaware that personal information is extracted

from public sites. In addition, the consumer may be unaware that data provided in an e-commerce transaction is being shared (lawfully) with data brokers who then package and sell it. The consumer may not know that personal data is at risk of breach. Social actors in general are unable to conceptualize the harms that result from data sharing, whether lawful or not. Risk, anxiety, and insecurity are slow to translate into legally cognizable damages.

CHAPTER FOUR

The Internet of Things and the Smart City

> The potential for cities to transform themselves for the better has never been greater, thanks to digital technologies that enable the exchange of data on a vast scale.
>
> —Promotional brochure for Siemens smart infrastructure

> The urban environment will be the locus where drivers of instability will converge.
>
> —Pentagon video on the future challenges posed by cities

From the IoT to the Smart City and Back

The ubiquity of technological bargains explored in the preceding chapter presents itself in one technological innovation in particular: the Internet of Things, or the IoT. This development can be seen by following David Lyon's division of computer history into three phases. First came the mainframe computer, which gave way to personal computing, which in turn ushered forth "ubiquitous" computing activities, "where computing machinery is embedded, more or less invisibly, in the environments of everyday life."[1] This last phase is marked by the rise of the IoT: "sensors and computation embedded within physical objects that then connect, communicate, and/or transmit information with or between each other through the Internet."[2] Here are some examples of nodes within the IoT: A web link to a camera trained on an eagle's nest allows Internet viewers to observe the behavior of the birds in real time. A heart monitor worn by a patient is linked to her cardiologist's office so that results can be observed and recorded over time. An athlete's watch records the body's reaction to a training regimen. Internet communications technology pervades more areas of social life, creating new networks and connecting those that exist already.

The life-world has become increasingly networked as (1) more of our activities are subject to monitoring and reporting (even by our own design, as with fitness watches), and (2) the monitoring of activities is interconnected by the firms offering IoT technologies and the sharing of data among those who collect it. As noted in the preceding chapter, it is simply more difficult to navigate the world we live in without encountering data collection points. Thus, the networked life-world is another phenomenon that contributes to the irresistibility of the surveillance society.

As Sadowski and Pasquale put it, "It no longer makes sense to think of 'the Internet' as a thing that one accesses via a computer. Not when the city itself is reimagined and reconstructed as a platform for and node within networked information communication technologies."[3] Here they are referencing the "smart city," in which governance is facilitated by ever-more-advanced systems that provide and monitor public services (e.g., security, "throughput" of traffic, health screening). Smart city initiatives are not perfectly co-extensive with the IoT—some functionalities do not depend on the Internet. Motion-activated gates and closed-circuit television (CCTV), for example, are not necessarily linked to the Internet and can operate without being connected. Nonetheless, the Internet has dramatically accelerated the use of automated technologies for governance, whether through data collection, policing, controlling access to public space, or other ways.

There are a few examples around the world of cities that have been *created* as "smart cities: New Songdo in Korea, for example, and Masdar in the United Arab Emirates. A city called PlanIT Valley outside of Paredes in Portugal was planned but never completed. More often, though, *existing* cities have been retrofitted with smart technologies.[4] One good example is Rio de Janeiro, Brazil: it is an old city that has adopted a rather comprehensive plan for executing government functions from a single control center. Thus, the number of people worldwide who live with smart infrastructure in their living environments far exceeds the residents of specially designed and new smart cities. In fact, it is unavoidable that all of us will come into contact with Internet-linked sensors if we spend any time at all in a city (or even on a major roadway).

But what are we to make of this networked environment of ubiquitous computing that has been constructed around us through government-industry partnerships? As numerous commentators point out, the cheerful and uncritical celebration of smart city initiatives often masks the harms to citizens. The ends to which this systems engineering is directed by government may conflict with the democratically determined ends that the public wants to prioritize. If systems expertise that helps to build the smart city comes from private firms, then city managers can be expected to rely on those firms, rather than majority preference, in making governance decisions. Obviously, the interests of system-designing firms and the good of the populace can often diverge.

In this linkage of the IoT with infrastructural city planning, the convenience and efficiency promised by the IoT draw the interest of local governments just as they do, on a smaller scale, individual consumers. Thus, government planners' attraction to the smart city idea reflects and reinforces an underlying "neoliberal ethos that prioritizes market-led and technological solutions to city governance and development, and it is perhaps no surprise that some of the strongest advocates for smart city development are big business."[5]

As one commentator expresses it, the goal of smart city planning is "to run the complicated infrastructure of a city with as little human intervention as possible."[6] In these governance arrangements, "machine politics will have a literal meaning—our interactions with the people and objects around us will be turned into data that computers in a control hub, not flesh-and-blood politicians, will analyze."[7] When the city is able to manage traffic flow, issue safety warnings, and limit access to public space through sensors reporting information to centrally linked information managers, certain logistical problems of governance can be solved, and efficiencies can be realized. Functions executed through smart technology range from mundane to disturbing:

- Trash collection
- Street repair

- Charging electric automobiles at public stations
- Traffic signals
- Public safety notification
- Noise monitoring
- Disabling cars when payments are late[8]
- Controlling access to public space
- Crowd control (using drones, for example).[9]

These IoT-driven governance solutions often come from private vendors who have the ability to harness Internet communications technology to the ends that city managers desire. As Sadowski and Pasquale point out, the structure of these incentives carries the risk that the interest of the vendors (i.e., making money) may come to supplant the interest of citizens (i.e., maintaining a workable and accountable government).

Of course, the smart city contains a range of phenomena, not all of which are prima facie threatening. In a 2011 lecture, Adam Greenfield offered a list similar to the one provided above, a taxonomy of networked "public objects" classified as objectionable or not according to whether they collect information or shape behavior.[10] At the "unobjectionable" end of the spectrum, he cited a traffic beacon that flashes to warn automobile drivers in oncoming cars that a cyclist or pedestrian is approaching from the opposite direction. This device reflects moment-to-moment changes in traffic conditions but collects no information, and it does not shape behavior aside from warning drivers to avoid collisions. At the other extreme, Greenfield classified as objectionable an interactive billboard that records passersby and registers their reaction to the billboard. The information collected holds value for advertisers, who now know how individuals (and groups of similar individuals) react to stimuli. Moreover, people might well change their behavior once they are singled out for not paying attention to the billboard. Thus, the interactive billboard both collects information and shapes behavior. Greenfield's taxonomy is an important step toward fine-grained critical appraisal of networked public life. The enthrallment commonly expressed in the face of these smart technologies misses the threats they introduce, but a blanket pessimism about all smart devices is uncritical and difficult to sustain as a coherent and cogent perspective. It is crucial to note that Greenfield is not in any sense opposed to technology for its own sake or in its essence. Instead, he has traced out the ways in which technologies built into the city's architecture and feeding data to government entities change not only the built environment of the city but also the experience of living in it. This is why his taxonomical distinctions pivoting on objectionable/unobjectionable are so important.

In one disturbing example that certainly lies on the far end of "objectionable," city police surveil public protests in Kiev, Ukraine.[11] The facial images are put through a recognition program to match faces with names, and then the individuals are notified that they have been registered with the government as protestors.[12] The chilling effect here on expression and association is almost too obvious to point out, but it is linked to smart technologies of governance.

The Intelligent Operations Center built in 2010 by IBM for the city of Rio de Janeiro is another case. It "draws together data streams from thirty agencies, including traffic and public transport, municipal and utility services, emergency services, weather feeds, and information sent in by employees and the public via phone, Internet and radio, into a single data analytics centre."[13] These are all familiar government functions; what is unusual is their consolidation into a "smart" operation. Thus, "with this NASAesque control room, the city of Rio is turned into a system for optimization and securitization."[14] In this configuration of government activity, efficiency is optimized, and the "data analytics centre" enables control of the population's movement and activity from one locus. The authors note that "the social contract is replaced by the corporate contract" in this instance. The people's consent to be governed in exchange for protection from their neighbors, which forms the basis for the social contract in liberal political theory, has been manipulated so that a different set of contractual bonds holds together the governance arrangements in the smart city. Government makes a "corporate contract" wherein the private corporation is paid to manage government services in a manner that is most efficient for government but not necessarily best for the citizens. This new contract is supported by "technocratic neoliberal ideology"[15] that constructs privatization as not only good but necessary, and encourages faith in technology to bring about the best possible living conditions. This vision of the city encourages "passive" citizenship and prizes convenience above all else.[16] Gary Marx maps this ideological phenomenon through his portrayal of "Rocky Bottoms."[17] Bottoms is Marx's creation. He is a fictitious security consultant who advocates surveillance technologies and scoffs at privacy-related objections to their use. Bottoms speaks in cobbled-together fragments of technocratic language that the author has taken from a range of real-life sources: popular culture references, ads for security products, and law enforcement literature, for example. He is a voice of technocratic optimism, and he is critical/suspicious of those who hesitate to embrace that optimism. His ideological intervention supports neoliberal projects such as the "smart city." While he is not an actually existing human, Bottoms speaks in phrases that have been uttered by actually existing people, and then recorded and reproduced. He illustrates how the work of belief-shaping vis-à-vis security-related technology happens, though it occurs outside the control of any one individual or group.

While the move toward "smart city" systems of governance might seem irresistible in view of the ideological pronouncements that support and justify it, closer questioning reveals divergences between public and government interests. For example, Sadowski and Pasquale ask, "Would autonomous car control systems prioritize preventing pedestrian deaths, or merely aspire to smooth flows of cars into and out of the city?"[18] At the most abstract level of bargaining, citizens consent to surrender their freedoms in exchange for protection. This is the social contract in its most basic form. But the basis for this agreement begins to deteriorate when government no longer seeks to promote safety above other ends. In order to reveal the stake that private corporations have in promoting and installing smart technologies in city infrastructure, it is first necessary to pierce the protective ideological shield that surrounds these privatized governing arrangements. This is not easy to do, as it requires one to buck the tide of faith in technology that equates human progress with technological advancement. Just as the Internet itself revolutionized the gathering of information and personal communication, the IoT has enabled us to act in the world more efficiently, whether through watching remote cameras via Internet feed or monitoring body systems. Technocratic rationality has become dominant, in part, because of the positive outcomes—in health, particularly—that it promotes. If the IoT can provide a better picture of cardiac health or a more effective exercise program, then it is unrealistic to expect people to oppose it. But the point is to understand the range of effects, whether they are bargained for or unanticipated. So it is worth noting that "the IoT is not simply a chance to watch people, but to produce and reproduce certain patterns of interaction."[19] Whether a fitness app on a "smart watch" was intended by its designer to produce "patterns of interaction" is somewhat beside the point: when operationalized it becomes a node in the IoT. Thus, the bodies of its users become conduction points for biopower, in Foucault's terms.[20] And each subject is a "vulnerable data subject."[21] As Sadowski and Pasquale put it, "close examination of the phenomenology of being a surveilled subject, a data subject, reveals the vulnerability of each resident of the smart city to extraction, oppression, and misrecognition."[22]

Further, the protocols of information control operating in the smart city disadvantage citizens at an even more basic level: they do not have access to the information on which governance decisions are based. There is a "deliberate exclusion" of the public, "an explicit statement that administrators are entitled to avail themselves of perspectives denied to the public."[23] In other words, it is not merely that citizens' interests come second to corporate interests, but also the citizens cannot determine what their reasoned judgment might be because the relevant information is denied them.

Engagement with the smart city appears irresistible through the combined intervention of state and industry, public and private actors. Surveillance is imposed and expanded through initiatives that further governance and profit. The trade-off involves a promise of optimization in exchange for this ceding of control over aspects of a city-dweller's life. However, Greenfield points out how "slippery" this use of "optimization" is.[24] "What is being claimed," he says, "is total in scope: that via these technologies, every register of urban life can be brought to an optimal state simultaneously and maintained in that state indefinitely, without cost."[25] This usage of the word *optimization* suggests that there are no trade-offs or compromises to be faced, and that perfection of all aspects of life can be attained without sacrificing anything in the process. We are all too familiar with the compromises made by legislatures and policy makers who seek solutions to a social problem. Passing the Clean Air Act, for example, imposed burdens on state governments as well as industry in exchange for reducing the pollutants in the air we breathe. The Clean Air Act sets limits on the amounts of each specific pollutant, for example, particulate matter. These limits are higher than zero, and that fact in itself shows the imperative of compromise in the legislative process. It would be better for us, of course, to breathe in *zero* particulates as opposed to the currently allowable amount under the Clean Air Act, but the bill passed because of the trade-offs between constituencies (industry, environmentalists, citizens, and local government)—including a higher-than-zero allowable limit of pollutants. And yet the advocates of the smart city promise that *all* the aspects of life in that environment will be optimized. The thickness and complexity of life in actual cities, together with the nature of the legislative process as seen at local state and federal levels, makes this claim of total optimization an "absurdity." Optimization of space and travel time, for instance, would shortchange such beloved activities as an evening stroll or sitting on the stoop of an apartment building.[26] Stoop-sitting is a less-than-optimal use of space, and aimless strolling a less-than-efficient use of time, but the cost of optimizing is to eliminate activities that make life worth living.

The claims of optimization often come from promoters of smart city development, and they insist all areas of life can be improved. Consider this language from the McKinsey Global Institute's report on smart cities: "Digital intelligence gives cities a fresh set of tools for improving the quality of life in practical, tangible ways. A wide range of applications now exists to make cities not only more efficient and productive but also more responsive and livable."[27]

In reality, some of these ends (efficiency, responsiveness, livability) conflict with each other. McKinsey's taxonomy of smart applications contains eight divisions: *Security/Healthcare/Mobility/Energy/Water/Waste/Economic Development* and *Housing/Engagement and Community*. Within and across these divisions, many of the specific applications have to do with *environmental quality*, for example:

- Real-time air quality information
- Infectious disease surveillance
- Data-based public health interventions: sanitations and hygiene
- Water consumption and tracking
- Leakage detection and control (water)
- Smart irrigation
- Water-quality monitoring
- Digital tracking and payment for waste disposal.[28]

All of these applications contribute potentially to environmental quality, which is a good whose value is obvious. "Sustainability" is an end well worth promoting by government. However, each of the applications listed above can be seen as *anti-human* in an important way.[29] The problems they are directed against are problems caused by humans, and the solutions to those problems involve more tracking, more monitoring, more data collection, and more surveillance. This point is not intended to undercut sustainability efforts or to discourage environmental awareness. Indeed, without such awareness (and perhaps even with it) the global environment is in grave danger. Rather, by showing how environmentalism and humanism can stand at odds in smart city planning, I mean merely to highlight the difficult choices to be made. All ends cannot be promoted at once, and optimizing one function can come at the cost of another. More environmental monitoring means more surveillance. That zero-sum proposition must be confronted so that the best choices can be made in a given specific situation.

In this brief exploration of the IoT's incorporation into the smart city, I have noted some of the built-in threats to democratic decision-making and to pursuit of the public good. These threats arise out of a transformation of the social contract to corporate contract wrought by government-industry collaborations in urban design. But there is also a problem of *scale* that arises when the IoT moves to the city level. At the individual body scale, we choose to use certain devices (a cell phone, a fitness watch) because we deem the resultant benefits to be worthwhile. I can even assess my physical fitness and say that my fitness watch has brought about a desired positive change. Similarly, at the scale of my home or living space, I can decide whether to use a smart thermostat or smart crib, and whether to continue using those tools in view of their effects on my life. All of the warnings, unknown consequences, and risks of data extraction described in the preceding pages are still there, but body-scale and home-scale IoT devices promise a specific convenience or efficiency. At the *city* scale, though, a new assumption is operationalized: that the city is a system (or machine) that can be totally administered and managed in the same way that a lawnmower or a central air conditioner might be. This claim—that a city is a system—raises the stakes of decisions to engage with technology because people become parts of that system. No longer simply individuals who opt for the bargain of using a watch or voice-activated phone (no matter how ill-advised such a choice might be), city residents are nodes that must function in the expected way or be singled out for attention and correction. I will have more to say about the specific problems related to treating cities as systems; here, I would simply like to identify the assumption for what it is: a mistake of scale. And before proceeding further along these lines, it is necessary to look genealogically at the governance techniques that enabled the move to city-as-system in the era of smart cities.

The "Re-Imposition of Old Chains": Precursors to the Smart City

However much the IoT may have accelerated the interpenetration of public and private actors in city governance in the smart city, we must bear in mind that deployment of technological innovations to private ends by city planners and developers did not begin in the Internet age. IoT technologies have simply furthered the irresistibility of a highly attractive partnership of political and economic elites seeking to consolidate power and profit, following a long-standing pattern. While the means afforded by smart technologies enable greater and more pervasive use of state power for private profit, the impulse to misdirect or maldistribute public resources is coeval with government itself. Excellent studies of nineteenth- and twentieth-century municipal governance have shown how "technologies"—both technological artifacts and effective, instrumental knowledge—have enabled individuals and firms to enrich themselves under the guise of providing good municipal government.[30] Gray Brechin's narrative of the municipal development of *Imperial San Francisco* is one brilliant study that focuses on that city's history in the late 1800s. City planning is entwined with political and economic projects by threads that constrain the city to grow and elaborate along the lines of a particular form of modern empire-building. Mining is at the center of this story, from the California Gold Rush of 1848 to the detonation of the atomic bomb in 1945. Mining companies and other firms enlist institutions of government in their enterprises, and in return they promise to the public vast enrichment and growth. The recursivity of this relationship is captured in the following formulation: "However elites may disagree among themselves even to the point of murder, they can all agree that the city *must* grow—and its land values rise—to assure the continuation of their dominion."[31] Some leading industrialists of the time were directly involved in mining, while others simply saw it as a way to facilitate growth that would benefit the regional and national elites. California's wealth helped the nation to engage in territorial expansion (primarily through war), and that expansion in turn provided new mining opportunities, thus closing the loop.

Brechin is strikingly adept at discovering and outlining, from the historical record of turn-of-the-century San Francisco, narratives of racial dominance, spectacular greed, and dynastic control by prominent individuals and families. The "thought shapers," as he terms them, were able to promote war and infrastructural development in much the same way that today's techno-evangelists urge the public to embrace technocratic solutions that often benefit government and private industry far more than they benefit the public at large. To take one example that mirrors the schemes deployed by boosters of the smart city, San Francisco elites were able to convince the public that damming the pristine Hetch Hetchy Valley, and pumping its water via aqueduct to the San Francisco area, would be in the public interest. In fact, as Brechin shows, the public realized little utility savings or other benefit from the project, but it did drive up land values so that speculators profited greatly.[32]

All in all, Brechin provides us with a comprehensive picture of municipal governance from the mid-nineteenth to early twentieth century in a key region of the United States. San Francisco is poised, by geography and demography, to play a pivotal role in national identity formation, wealth extraction, and geostrategic policy initiatives. The comparison with Rome and other European cities is striking, as the built environment brings San Francisco renown while the political machinations of the city's elite extend its influence south into Mexico and even across the Pacific. The public good is determined by the elites—such as senator James Phelan and William Randolph Hearst—so that it is synonymous with private enrichment and maintaining the status quo of power relations.

In chronological terms, Victor Valle's *City of Industry* picks up where *Imperial San Francisco* left off: the early to mid-twentieth century.[33] The focus of Valle's work is California's City of Industry, and the author provides a Foucault-inflected study that pinpoints legal and political technologies that enabled business owners to create a municipal entity (City of Industry, hereinafter "Industry"), and then to feed off that entity for decades through redevelopment projects, no-bid contracts, and money laundering, among other tools. Valle's intricately detailed and meticulously documented study is deeply disturbing in a number of ways, not least of which is by showing that the technologies themselves (such as the redevelopment laws and cooptation of local bureaucracy) were *by design* highly suitable to private enrichment: particularly opening up investment opportunities for developers and construction firms.

This malfeasance was present in the City of Industry from the very beginning—that is, from the moment it incorporated as a municipality. In order for Industry to incorporate, there had to be a majority of residents willing to vote for incorporation, and therefore it became necessary to keep residential numbers low. After all, the real "residents" of Industry were *industries:* the railroad, the mill, and the export firms. Corporate interests would likely clash with residents' interests, and so it would not do to have a populous and restive citizenry. In response, Industry's founders followed an incorporation strategy that bootstrapped the founding of a city onto the capitalistic designs of a small group of investors. It was a city without a constituency, or perhaps a city with a constituency that had no public interests, but only private ones. Meanwhile, the neighboring municipality of La Puente bore the brunt of Industry's externalized harms in terms of pollution, industrial blight, and lack of infrastructure.

Valle's description of the use of this shrewd and adept maneuvering in City of Industry recalls another pathbreaking study of municipal governance gone wrong: Robert Caro's famous portrayal of Robert Moses and the development of roads, bridges, and parks in New York City and Long Island.[34] Moses aspired to be the world's greatest builder, and to that end he developed mechanisms for garnering public funding for construction projects around the New York region. Though he never held public office, he managed to develop and utilize mechanisms that enabled both the completion of specific projects and preservation of his own power base over the long term. From the 1920s to the 1960s he was able to operate more or less unchallenged in these endeavors. Moses utilized a variety of political and legal techniques, including inventive statutory drafting and work-camp–like labor policies. Then end result was a regime-within-a-city that directed and controlled development according to an unelected and unaccountable official's vision rather than the democratically determined public good.

Brechin, Valle, and Caro provide numerous examples of city planning and development initiatives that relied on then-new forms of technology and promised certain benefits to the public. As the authors show, however, those initiatives gave the public far less than promised, and in fact inured to the benefit of financial elites and other select elements of the citizenry. In Robert Moses's park- and road-building projects, Caro exposes the deliberately segregationist aims that privileged white suburbanites at the expense of African American and Latino residents of the five boroughs.[35] To the extent that any bargain was struck in order to gain support or approval for such projects, it was a bargain that benefited a select few and shut out many others. The point is that seemingly universal solutions that actually carry significant disadvantages for many of us are by no means new. The "neophilia"[36] that fuels smart city proliferation today was present at these earlier moments as well and provided, then as now, an independent basis for the public to support new ventures, aside from the specific benefits in convenience or efficiency that were promised. Today, data-gathering corporations, purveyors of smart technology, and government elements seeking political control all benefit from the creation of smart cities and the retrofitting of existing urbanscapes with smart technology nodes and functions. In a way, these covert governance projects are more insidious than their earlier counterparts because they are more likely to escape notice. The Hetch Hetchy water project's backers were open about their support, and the underwhelming public benefits were plain to see as soon as the project became operational.

One of the observations to be gleaned from a brief review of earlier municipal corruption is that technology alone is not responsible for the configurations of state power concentrated on surveillance that we see in the smart city today. "New" surveillance allows state actors (and private actors) to do more with greater efficiency than before, but the design and implementation invest those technologies with a political valence, as Langdon Winner has shown.[37] The genealogy presented illuminates linkages between technology and politics.

Viewed against this historical/genealogical background, the smart city movement is less surprising and radical. It dresses up long-existing power relations, showing us that "the sensors of the smart city will amount to little more than a technologized reimposition of old chains."[38] Ironically, this "re-imposition of old chains" is a historical pattern that recurs while smart cities' advocates present each smart cities as history-less. Greenfield observes: "They don't have to reckon with the messy accumulations of history, with existing neighborhoods and the claims and constituencies they inevitably give rise to, with the densely tangled ways of doing and being that make any real place what it is."[39]

Smart cities exist in "generic time" and break with the past in favor of a "proximate future" that is just around the corner but never actually arrives.[40] Conceptualized this way, they obscure the specific histories of the places they occupy. These histories comprise unjust enrichment of city developers and managers (negative and lamentable, yet instructive) as well as culturally rich traditions that sustain communities over time (salutary, grounding, and unifying). In fact, not only the "time" in which smart cities are imagined to exist, but also technologies themselves that make it up, are generic, without qualities. They simply make life better but are not specific to time, place, cultural setting, language, climate, or anything else.[41] In fact, the environment itself is mere "backdrop," a potential source of "friction" to overcome.[42]

Sadowski and Pasquale point out that the Foucauldian paradigm of political control from the top down, conceptualized as panoptic control in which citizens are always under scrutiny by the government machine, is somewhat less accurate in today's social settings.[43] To some extent, Deleuze and Guattari's rhizomatic metaphor, in which power moves through a system of ever-branching roots (similar to a computer network), is more fully descriptive.[44] Nayar seconds this observation.[45] The preference for rhizome over panopticon as descriptive device is well-founded in the sense that so many micro-relationships extend out from a surveillance-defined social order. The state watches citizens and gathers data; private firms gather data for marketing purposes, and other private firms (e.g., telecoms) assist government in surveilling. All of this is still top-down, and yet surveillance is also co-produced by citizens who surveil each other ("If you see something, say something") and surveil themselves as well—by using loyalty cards, for example. Nayar calls this "co-production" of surveillance.[46] Simultaneously, individuals are separated into discrete parts—they become *di*viduals—as classificatory processes fragment them, and this further complicates the splintering and branching of surveillance activities. But, while manifold surveillance relationships comprise a mosaic of interactions and linkages, genealogies of power are still important and worth tracing out. Understanding how private/public linkages, such as those depicted by Brechin, Valle, and Caro, have long perverted and corrupted governance is vital. This awareness serves as a corrective to the uncritical celebration of smart technology that we see so often in popular discourse. Gary Marx points out the fallacy of the 100 percent fail-safe as one set of beliefs accompanying technocratic rationality: it keeps people hoping and searching for the perfect design that will eliminate all possibilities of harm.[47]

Tracing the development of legal and political technologies also serves to underscore the ever-present element of discretion in governance. Whether we look at San Francisco in the 1890s, New York in the 1930s, City of Industry in the 1960s, or Rio de Janeiro today, political decision makers are always invested with discretion. The choice to use smart technologies in the first place, and the choice of which ones to develop and emphasize, involves the use of discretion. That element of governance is inescapable—and by juxtaposing the governance patters of today with those of earlier periods, we see that the citizenry can always fall victim to misdirection of government resources. The processes have become more insidious and in some ways harder to notice, but the impulse to misdirection remains the same.

Bargaining in the Smart City

As I noted earlier, the choice to live in a "smart city" rarely presents itself as such. Individuals consciously adopt IoT technologies on a micro-scale, in the activities of their daily lives: Alexa answers their questions; a computer program regulates the heating of their homes; a fitness watch records biometric data during exercise; a "smart crib" takes care of the baby.[48] Maybe we are aware of the costs of convenience in each scenario and maybe we aren't, but either way, we consciously make a choice to buy the Internet-connected object or service (or not to buy it). We choose to carry about or install these items, and they are "at hand," structuring our immediate environment; we choose to interact with them.

Can the same be said of the smart city, in which these elements of the environment are "writ large"? The line between individual "tools" and the ubiquitous "presence" of computing processes, as David Lyon phrases this distinction, can be blurry, but we gain a bit of clarity by looking outside the home to the city.[49] We don't choose every feature of the environment of our city, suburb, or exurb. Of course, some people move to a city deliberately who were not born there. And voters can often reject a proposed new program. But there will always be some features of the cityscape that we did not choose, whether because they were adopted without our consent, preceded our arrival, or simply escaped our notice. Since there are relatively few places in the world, at present, where a smart city was built from nothing, on previously uninhabited land, very few city dwellers bought into the smart city idea wholesale. Instead of bargaining for a spot in a smart city, it is overwhelmingly more likely that a smart city grew up around us.[50]

While there is limited opportunity for *individuals* to bargain in the process of creating a smart city or converting an existing city to smart functionality, there is nonetheless other bargaining going on. As mentioned earlier, commentators have noted that the social contract has been replaced by the "corporate contract." Government entities decide on the price at which smart technologies are worth buying, thereby making a contract between the state and Siemens, or IBM. This contract specifies what each side will give up and what each side will get. In the language of contract law, we might say the public are third-party beneficiaries of the corporate contract. In other words, the contract creates a benefit for the public, but they are not parties to its negotiation or dissolution and are therefore unable to influence its content. Of course, the value of this "beneficiary" status is highly variable, and citizens are just as likely to experience externalized harms as incidental benefits, as the next section of this chapter will explore.

Illusory Bargains and Unfulfillable Promises

There are essentially two sets of concerns voiced in the critical literature on smart cities that come into particularly clear focus when viewed in relation to the idea of bargaining. The first concern is that smart city developers and their supporters in government cannot actually deliver the city that they promise because the specifics of the promise are in fact impossible to fulfill. The smart city—whether it is a new entity entirely or a retrofitted already existing city—is presented not merely as an improvement of particular municipal functions but as a total "optimization" of city life, as noted briefly above. But this total optimization is impossible to achieve as well as difficult to measure, precisely because it is an optimization of *everything*. Total optimization will come to fruition in the "proximate future" (not yet here but very close)[51] and will deliver the city's residents and managers a place that has been freed from all of the risks, inefficiencies, and unanticipated occurrences that have been experienced throughout human history. When unpacked, this promise is clearly seen for what it is: an impossibility. Total optimization would require perfect knowledge, as well as freedom from the contingency that has always been a central fact of the human condition. Second, as discussed briefly in the previous section but more fully below, the bargains premised on this total optimization are illusory because they are usually designed not for optimization of the quality of residents' lives, but rather to ends that benefit government and developers. In fact, some ends come in conflict with each other, such as environmental protection and freedom from monitoring. The benefits to city residents, then, do not materialize because they are not really what the smart artifacts are designed for; thus, the putative bargain is illusory as well as impossible to deliver. These two related concerns—impossibility and illusoriness—are the focus of the remainder of this chapter.

One unfulfillable promise is that smart cities will bring about a risk-free future. To be precise, the vision of a risk-free (and uncertainty-free) future rests on the assumption that "the world is in principle perfectly knowable, its contents enumerable and their relations capable of being meaningfully encoded in the state of a technical system, without bias or distortion."[52] Perfect knowledge would anticipate and plan for every contingency. It would include all that there is to know, including the future events comprising every complex causal chain. It would take into account all the consequences flowing from the deployment of every piece of smart technology. All human reactions and responses would be planned for. Even though complete and perfect knowledge can never be attained, it has been a fantasy pursued by many thinkers through the ages. To embrace this technological positivism, one would have to discard the lessons we have learned through history about human unpredictability and unintended consequences. We would be obliged to eschew Hannah Arendt's warning that humans are capable of the "infinitely improbable" and to treat them instead as parts of a system, a machine bound by the functionalities built into its design.[53] But human societies are not systems in the sense of a set of components that always behave in lawlike and predictable ways. Humans often act irrationally, impelled by conflicting desires and complex motives. They form attachments to ideas and to other people, and those attachments cannot always be anticipated. As Lyon reminds us, people imagine the world, projecting their fears and wishes and impressions into a social imaginary that mediates their actions in the realm of social relations.[54] The experience of visiting or revisiting a city, for instance, is shaped for each of us by what we imagine or expect the city to be. These aspects of human being-in-the-world have long posed a fundamental dilemma for social scientists in that the subject-matter they study differs in an important way from the subject-matter of the physical sciences. Why do revolutions occur in one case and not another? How do leaders defy predictions and rise to power despite seemingly insurmountable obstacles?

The very same complexity of human experience that challenges social scientists also confounds technocrats who see the city as a system and offer governance solutions accordingly.

This overly optimistic yet short-sighted hubris arises in connection with other problems of governance as well. The fallacy of the perfect technocratic solution has led, for example, to the belief that the risk of terror attacks can be eliminated completely if only the right antiterror research-and-development projects are commissioned, the right defensive weapons designed. In reality, government security agencies have not been able to eliminate the risk of terrorist acts entirely. They *have* been able to respond to specific threats, adjust search protocols, and direct preventative force against known threats such as airplane hijacking and shoe bombs, for example. Of course, these efforts have saved lives and thwarted potential attacks, and they are necessary. My point is only that it is impossible to know in advance all the plans that a person or group bent on mass violence might hatch, just as it is impossible to anticipate all infrastructural failures or weaknesses that might be exposed by human action or even by environmental catastrophe. While it is necessary to guard against known threats, the risk of future terror attacks cannot be reduced to zero. In a critical analysis of technocratic ideology, Gary Marx lists a number of "information-age techno-fallacies" that result from this belief in a risk-free future: the fallacy of the "sure shot" and the "100% fail-safe" are two such fallacies.[55] The "sure shot" is an alluring claim, but it has never been realized. To quote Greenfield again, "no globally optimized solution to the city can ever be found."[56]

And yet planners, developers, government officials, and some journalists persist in what Torin Monahan calls "algorithmic fetishism."[57] Algorithms will succeed where other instrumental designs have failed (so the promise goes), and they will bring about a perfectly planned and totally administered governmental response to urban problems. Monahan has compared "solutions" developed, at the behest of IBM in its "Smarter Cities Challenge," for more effective governance of US cities. In response to this challenge, designers created plans for Durham, North Carolina; Pittsburgh, Pennsylvania; and St. Louis, Missouri. He notes that predictive analytics were included in those plans that aimed to address crime prevention, high school truancy, and other social problems, and that algorithms repeatedly take the place of funded social services in this version of social-service planning.[58] Residents need services, but they get data collection instead.

The plans for each of these cities involve data-driven solutions instead of monetary assistance, and this approach is consistent with a neoliberal ethos: privatized responses to urban problems emphasizing top-down regulation and displaying an insensitivity to vulnerable and underserved populations. Of course, these efforts are not politically neutral: "algorithms cannot be separated from the context of their production and use."[59] In the Smarter Cities Challenge, part of that context is neoliberalism itself, and so the planners started from fundamentally neoliberal assumptions: redistribution of resources is not warranted, government should not increase its role in ensuring citizens' well-being, and experts should manage cities with minimal citizen input.

Similarly, Rob Kitchin specifies some of the particulars that are overlooked and ignored in municipal governments' haste to adopt smart functionalities:

> Technocratic forms of governance are highly narrow in scope and reductionist and functionalist in approach, based on a limited set of particular kinds of data and failing to take account of the wider effects of culture, politics, policy, governance and capital that shape city life and how it unfolds. Technological solutions on their own are not going to solve the deep rooted structural problems in cities as they do not address their root causes. Rather they only enable the more efficient management of the manifestations of those problems.[60]

Incarceration, welfare eligibility restrictions, and expansion of criminal justice-related surveillance are familiar enough to observers of the woes of post-industrial US cities; Kitchin and Monahan contribute the insight that data-driven solutions leave the problems themselves intact while actually making day-to-day life harder for urban residents.[61] One wonders how governance might proceed if techno-utopian fantasies were abandoned in favor of a realistic pluralism that incorporated input from residents about the mundane details of daily life and the concerns that interfered with their enjoyment of a livable city. An inclusive conversation would start with the acknowledgment that there is no "sure shot"—so that policy makers could focus on more limited, yet feasible, responses. Local democracy is not a new idea, of course, but it is eclipsed all too frequently by the seductive claims made by techno-evangelists promising total optimization.

In addition to being unfulfillable, promises of smart governance are illusory because they are not in truth primarily directed at the ends they claim to serve. The true aim is hidden. In some instances, the veil is pulled back so that there is not even the pretense of concern for residents' well-being, or openness to residents' input. Greenfield notes that the promotional literature wooing city planners frequently fails to mention residents as stakeholders in the planned smart city.[62] The absence of appeals to the people who would actually live in these spaces tells us where residents rank in relation to other constituencies: near the bottom of the list, if not ignored entirely.

And so deployment of smart functionalities works to strengthen government control. According to Kitchin, successive interventions in city government create a "corporate path dependency"[63] as one smart functionality facilitates another. This trend can be seen in IBM's Intelligent Operations Center in Rio de Janeiro, where more and more functions are managed from the same control center. If centralization and coordination are the goals, it only makes sense to operate multiple projects out of a single center. And if services are already being purchased from a vendor, it is much easier and more efficient to continue purchasing new ones from that same vendor, in much the same way that Pasquale describes consumers tracking into a groove of purchasing patterns by utilizing the same platform. In both cases, past actions pave the way for future ones.

"The smart cities of the future," Kitchin warns, "will likely reflect narrow corporate and state visions, rather than the desires of wider society."[64] Thus, the invitations (direct and indirect) for people to support and live in smart cities amount to illusory bargains because the real aims are corporate profit and increased state power. To put it another way, rather than expecting benefits to flow to the public, we ought to be aware of the "damage this technics of watchfulness would inflict on a functioning place."[65] The functioning of a city encompasses a range of individual and collective activities: competition over resources; historical claims of justice, complaint, and protest; flows of people in and out; and the disproportionate growth of city sectors, to give a few examples. From the perspective of city-as-system, all of these activities appear as nothing more than sources of "friction" to be eliminated, in the manner of a poorly oiled pulley that slows the movement of a machine's belts for lack of lubrication.[66] Human behavior and human diversity are reduced to unwanted friction. And the place where interaction takes place has been phased out as well. The public sphere would be—has been—a space for residents to encounter one another, exchange ideas, and pursue individual or collective goals, but those encounters are contingent and unpredictable and therefore do not fit with the design of a totally administered city.[67]

Conclusion

This chapter has explored the "ubiquitous computing" of the IoT and its growth into the built environment of the infrastructure of cities in the United States and elsewhere. The claimed efficiency of technological solutions, the allure of the new (neophilia), and the overall frequency with which technological artifacts are encountered all combine to make the IoT and the smart cities it constructs appear irresistible. We cannot resist the bargain of the smart city, it seems, and after all, why should we? We adopt IoT solutions on the individual scale on an apparently voluntary basis, and the city is nothing more than an extension of that choice.

But things are more complicated. First, the move from personal convenience to city-level planning is a change in scale. The critical literature on smart cities reveals the statist and corporatist assumptions driving the design and marketing of smart city technologies. One root of the problem, according to Greenfield, is a "semantic contamination." This chapter has referenced the variety and unpredictability of the everyday lives led by city dwellers for as long as there have been cities. A city has its own history, its own distinctive features, and its own well-known and time-honored pursuits, such as strolling through an old neighborhood—none of which factor into efficiency calculations in general or the smart city business model in particular. Thus, "an idea endemic to the culture of business administration has effectively been copy-and-pasted into a realm where it has no place and makes no sense."[68] Viewed this way, the entire enterprise of building a smart city from IoT devices seems misguided. Bearing in mind that technology is neither good nor bad in itself (a claim Greenfield agrees with), one could nonetheless find fault with the shared assumptions of smart city developers:

- That risk and contingency can be eliminated entirely.
- That knowledge can be perfected.

- That optimization of one aspect of city life can always be attained without diminishing another.
- That efficient management is to be valued above all else.
- That the concerns of residents do not factor into the planning process.

These are certainly assumptions worth contesting, worth resisting. Whether or not people will expend the effort required to resist them is another question.

A Pentagon video produced a few years ago offers a deeply disturbing vision of the technologized cities of the future.[69] The film is intended to persuade viewers of the need for a new kind of warfare that pacifies cities. Cities will be the battlegrounds of the future, the narrator warns, and it will be necessary to fight differently there as compared to the way wars have previously been fought. Urban populations must necessarily be controlled, partly because a small number of people can inflict casualties far out of proportion to their numbers when they attack the highly concentrated population of a city. The inescapable conclusion from this line of argument is that technologized solutions are the key to survival (of the human race?) and that those solutions involve force as well as persuasion. The military becomes responsible for governance, and governance takes on a military orientation.

But this short film on "megacities" does not operate solely at the level of rational argument, even though the narrator makes use of statistical evidence. Video, as compared to the printed word, evokes visceral reactions more quickly. Burning buildings and injured bodies, crowded slums, and gangs of armed rebels appear in constant succession, leaving the viewer unsettled and even traumatized. I have to say that I experienced those effects even though I was expecting the content. Certainly the film's message is conveyed on the rational and emotional levels simultaneously. The statistics become more powerful when accompanied by fear- and dread-inducing images. No one want to stare into a row of gun barrels or huddle in the corner of a high-rise slum apartment engulfed in flames. Emotionally, the video pulls for a reaction of surrender and abandonment: let the military fight these urban battles on behalf of the peaceful citizenry with sophisticated weapons and advanced methods of persuasion.

The problem, of course, is that there is no clear line between the enemy and the rest of the public. All of the city residents are pacified at once; the battle is for passivity and docility throughout the population. Anyone in the civilian population could be the enemy, but the military forces stand unquestionability on the side of right. Nowhere in the film do we hear the faintest mention of resource allocation, protection of human rights, or citizens' role in government. Thus, this promotional video fits perfectly with the ideology of the smart city. Cities are to be managed, by force if necessary; efficiency and orderliness are prioritized, and human unpredictability, spontaneity, and ties to tradition are all viewed as suspect. The Pentagon video is a stark representation of the ideology of the smart city.

As I contended earlier, individuals enjoy more range of choice when interacting with the IoT devices they purchase, rent, or otherwise control. Even if they are surrendering personal data to a greater degree than they realize, they are deciding whether to engage with those devices. At the scale of the city, though, there is less choice involved. More often than not, residents of the smart city were there before smart technologies were introduced, and so no bargain was offered them. One exception might be gated communities, which residents *do* choose, but those communities may or may not deploy smart technologies.[70] Thus, the bargain struck to create a smart city or import smart technologies into an existing city does not involve the consent of its residents. As this chapter has shown, the lack of citizen consent or input works to their detriment. Smart cities turn out to be no bargain at all for their residents or for the history- and tradition-linked cities themselves. The historical city is supplanted by a city that is generic both in space and time.

The following chapter will explore some avenues of resistance to surveillance and control that the state attempts through smart technologies. After that, the concluding chapter will explore the larger-scale bargain structuring modern societies: the social contract. Whether we imagine the social contract as a written constitution or a fictitious moment in the development of the modern society, it remains relevant today as a structuring force in contemporary democratic political orders. It is the legitimating element in those governments, and without it the state must resort to force to preserve itself. Thus, it is appropriate to ask what effect the smaller-scale bargains described in the preceding pages have produced on the larger contract that underpins constitutional democracies here in the United States and elsewhere. Smart cities are an important mediating link between the political order and citizens of the nation. Citizens experience government most directly in the cities (and suburbs) where they live, but smart cities and the private entities that help to run them are allowed to exist in the first place by virtue of a national government that privileges technocratic, neoliberal solutions in general.

CHAPTER FIVE

Avenues of Resistance

The core attribute of Bitcoin is that it allows unrelated individuals and organizations to have confidence in transactions without trusting intermediaries or a legal system.

—Kevin Werbach and Nicholas Cornell, "Contracts Ex Machina"

Further removed from a sentry role is an app called iSpy, which can be installed on an iPad or iPhone for a nominal sum. It allows the user to patch into the feeds from thousands of surveillance cameras all over the world.

—Richard W. Evans, "'The Footage Is Decisive': Applying the Thinking of Marshall McLuhan to CCTV and Police Misconduct"

The preceding discussion has detailed the ways in which surveillance and "dataveillance" create privacy harms that are both widespread and hard to detect. State surveillance relies, inter alia, on captured facial images, closed-circuit camera footage, physical searches, X-ray machines, and also large-scale data collection that sweeps up social media artifacts from millions of users, as we learned from Edward Snowden's revelations. The dataveillance of social media that Snowden uncovered was possible in the first place because of the cooperation of state and corporate actors, as platform owners such as Facebook allowed the NSA to access user data.

But corporations also collect data for their own purposes, as was discussed in chapters 2 and 3: to market their products more effectively, and in some cases, to sell the data to third parties. Taken together, all of these activities of watching create a web that entangles us at multiple points as we go about living our lives. In view of this predicament that people face in their life experiences as consumers and citizens, I have termed the everyday encounters with surveillance infrastructures *irresistible bargains*.[1] Of course, the irresistibility here is not absolute but probabilistic. Lack of awareness, the force of inertia, failure of prudence, and the allure of short-term gain create obstacles to resistance, and also result in a greater likelihood that people will not resist. As a matter of probability, then, many or most actors will go along, making bargains that result in loss of control over personal information, notwithstanding how lopsided or ultimately disadvantageous those bargains might be.

But to say that *no one* can refuse or resist the technological bargains that are continually on offer would fly in the face of reality. Resistance options are available, and while it is worth asking why those options are not explored more often, it is still necessary to inventory the types of resistance that actually occur. This chapter will list and briefly describe five resistance options: *countersurveillance, encryption, cryptocurrency, cash transactions,* and *surveillance art*.

Countersurveillance

Countersurveillance involves " turning the tables and surveilling those who are doing the surveillance."[2] This phenomenon became widely known after the emergence of a 1991 videotape in which Los Angeles police officers could be seen beating and kicking African American civilian Rodney King during an arrest. The footage was captured by a private citizen with a handheld videotape recorder, and it provided key evidence in the two criminal trials—state and federal—of the officers involved.[3] As recording technologies advanced over the ensuing three decades, the number of occasions where citizens filmed police/civilian conflicts increased dramatically. Today, an ordinary cell phone can produce high-quality still and video images, and most people carry a cell phone at all times. Consequently, many police use-of-force incidents have been captured on bystanders' cell phones, and the footage is routinely used at trial when a case is prosecuted—in fact, the existence of such evidence has in some cases made it impossible to avoid, at the least, further investigation.[4] Since the images can be circulated quickly and without cost through the Internet, their impact is not limited to influencing legal proceedings: they shape public opinion as well.

These incidental uses of video recording devices to uncover police misconduct led over time to the deployment, by police departments, of cameras mounted on the dashboards of police cruisers and on officers' uniforms. Dashcams and bodycams have become mandatory in some jurisdictions. They can be termed countersurveillance in the sense described above, as they involve turning surveillance technology against the state,[5] but they are distinguishable from civilian-made video recordings in that they are deployed by the state to surveil itself. In a functional sense, though, dashcams and bodycams work to counter the state's efforts to control the citizenry. Now, the actions of the police are on record alongside the actions of the public. The obvious value of such initiatives lies in their ability to provide evidence of all police/civilian interactions, and also in their tendency to deter police misconduct in the first place by making police behavior public. Of course, this requires that the cameras are operating correctly; they cannot document anything if they are turned off. Thus, oversight at another level is needed in order for the cameras to serve their ostensible purpose.

There is another concern associated with police cameras that is, perhaps, an unintended consequence. Dashcams and bodycams create a record of two types of information at once: how the police behave and how people in policed communities behave. Constantly running cameras compound the scrutiny of residents' lives that is already happening by virtue of the police being there and observing residents visually. The cameras create a record that includes information that the officers did not deem worthy of investigation at the time. For those who (rightly) feel that observation by the state tends inevitably to bring problems, there is greater cause for concern when more data is being collected. I mention these concerns not to argue against the use of dashcams and bodycams but to highlight the complexity of countersurveillance measures. In the surveillance society, multiple processes of watching occur simultaneously, and it is therefore difficult for individuals to avoid becoming the object of scrutiny. As the means of surveillance become more sophisticated, they also tend cumulatively to generate more data.

One commentator has used case studies involving Australian police departments to highlight the difficulty of holding the state to account for misconduct occurring under the camera's eye. In the cases discussed, the state can be seen to vacillate between *techno-positivism* and *techno-skepticism* in order to escape blame. If the camera does not show a particular aspect of a given incident, it cannot be taken as fact, no matter what other evidence tends to suggest that it occurred. This *positivistic* stance essentially says that everything worth knowing can be shown by the camera. Whatever else a civilian might allege did not, for legal purposes, occur. If the camera did not see it, it never happened. Thus, the police escape responsibility for any unrecorded misconduct. But when the camera *does* show apparent misconduct—excessive force, for example—the official claim is one of *skepticism:* the camera does not tell the whole story. And once again, the claim of misconduct fails.[6] The techno-skepticist position was deployed during the trial of the officers who beat Rodney King. Seemingly indisputable evidence was explained away, in the courtroom, by the suggestion that the footage was somehow misleading. Of course, when taken together, positivism and skepticism cover all cases, when the conduct was caught on film and when the camera missed it.

Like police use of dashcams and bodycams, home security systems amount to ambivalent surveillance. On one hand, these systems—with cameras that record intruders—enable individuals to engage in counter-surveillance by watching others. What began with simple security cameras has now expanded in scope and sophistication to include such services as Amazon Ring, which allows homeowners to view human activity on or near their property through a doorbell camera.[7] Customers can view the feed remotely, and it can also be shared with police departments. Four hundred departments nationwide have connected with this Amazon-owned program, thereby gaining access to the footage of individual cameras within a given jurisdiction. Thus, the homeowner's property itself, as well as the activities of outsiders, is open to scrutiny by the state once the doorbell camera feed has been accessed by the police. While customers can decline to link their doorbell cams to the police department, there is a social pressure here that is also seen in other instances of "voluntary" cooperation with law enforcement. If one opts out of sharing the doorbell cam footage with municipal police, that refusal can appear suspicious, as if the resident might have something to hide. Once again, a countersurveillance tool can end up providing data to the government even as that tool generates data for the user.

Crowdsourced crime reporting apps are an additional variety of countersurveillance activity that has emerged in recent years. They allow the public access to police emergency calls, as amateur radio operators have long been able to get, and they make the information flow far more user-friendly than raw police-band chatter. Crime apps such as the recently developed Citizen "convert raw scanner traffic—which is by nature unvetted and mostly operational—into filtered curated *digital content,* legible to regular people, rendered on a map in a far more digestible form" (emphasis in original).[8] The app shows a dot indicating the user's location, as well as a graphic of real-time crime events in the vicinity. Citizen replaced Vigilante, an earlier app created by the same developer. Vigilante was discontinued amid concerns that it encouraged violence, as users took enforcement and retaliation into their own hands. Having the most current information about safety threats is obviously valuable. Even so, the negative potential of these services is clear: they tend to assign blame for criminal conduct without any fact-finding or investigation, based only on an initial report. And at worst, they encourage vigilantism.

Encryption

Encryption thwarts surveillance in a different way from video recording countersurveillance: rather than uncovering government misconduct directly, it allows private users to communicate without fear of detection or interception. In the context of contemporary privacy protection, encryption is a method of obscuring contents of electronic mail and other digital communications by "wrap[ping]" them "in a protective shield, which is a code composed of hundreds, or even thousands, of random numbers and case-sensitive letters."[9] It is vastly more difficult to break an encrypted communication as compared to a simple password because of the number of characters utilized by an encryption program. Thus, encryption is a popular and effective way of preserving secrecy in communication, and it is used by activists, journalists, and government officials, among others. By virtue of the difficulty in breaking through it, encrypted communication provides an assurance to its users that their messages cannot be read by anyone other than the intended recipient.

In order for encrypted communication to take place, both parties to the conversation must be using an encryption program. The most commonly used program is PGP, or "pretty good privacy." This program itself is hardly a secret: tutorials on using it can be found on the Internet. Sender and recipient are linked through the use of a pair of keys: a *public* key for encryption on the sending end and a *private* key for decryption on the receiving end. If a sender (S) knows the recipient (R)'s public key, then S can encrypt and send a message to R. However, only R can open (decrypt) the message because R has a private key that R alone knows, and only that private key can decrypt the message.

When Edward Snowden initially contacted Glenn Greenwald, the journalist who eventually broke Snowden's story, Snowden used a pseudonym to email him. Snowden urged Greenwald to use encrypted emails when they communicated, and in fact he looked first, online, for Greenwald's "public key" in order to begin encrypted communications. When Snowden could not find the public key, he assumed that Greenwald was not using PGP encryption. In fact, Snowden waited until Greenwald had begun using encryption before forwarding him any information.[10] It almost goes without saying that the importance of encryption as a means of resisting surveillance was demonstrated best in the biggest whistleblower case ever known: Snowden's disclosure of massive surveillance of email and social media users under the NSA's PRISM and other initiatives.[11] Encryption was necessary for the leak to occur and for the public to learn of it through reporting by Greenwald and others. Greenwald himself acknowledges that he nearly missed reporting on "one of the largest and most consequential national security leaks in U.S. history" simply because he waited, at first, to learn the encryption process.[12]

The availability of encryption protocols increased dramatically around 2015 when Apple made them standard on the iPhone.[13] It was not simply that encryption (for data storage, not email) was available to most if not all cell phone users; even more significant was the convenience offered by embedding encryption technology in the cell phone itself. People who might not make the extra effort to learn about it or install it would have data encryption at their fingertips without having to expend any effort. Of course, this convenience applied to iMessage and other social media services but not to email. As explained above, encryption requires the linking of two keys between sender and recipient. Thus, senders cannot always encrypt their outgoing email since some recipients lack the capability to decrypt.[14] For those who are able to use it, encryption provides an assurance of privacy in electronic communications even in a time of pervasive and sophisticated surveillance.

Cryptocurrency

The emergence of *cryptocurrency* expands the use of encryption even further, as transactions using cryptocurrency are themselves encrypted so that only the transacting parties can read them. Cryptocurrency is "digital currency not issued by any bank or sovereign state."[15]

We are accustomed to thinking about currency as a symbolic medium of exchange that works because a government has backed or guaranteed its value. As a result of that guarantee, we know that others will recognize the value of the currency as long as the government remains in existence. Cryptocurrency does not enjoy government backing, and so it requires some other mechanism of trust in order to function as a medium of exchange. In the case of Bitcoin, the most popular and widespread cryptocurrency, that medium is the blockchain. The blockchain supports Bitcoin by offering a chain of blocks on which transactions are encoded.[16] Individuals who transact using Bitcoin memorialize their transactions on a chain of blocks that grows when new transactions are added to it.[17] Users of the blockchain can track their transactions by means of one of the many ledgers available, but there is no master ledger. The absence of a master, or original, ledger is one of the features that preserves collective public control over the system while at the same time providing an accurate accounting of each party's activity. As Kevin Werbach puts it, "any participant in the network can maintain an instantiation of the ledger, yet be confident it matches all the others."[18]

Contracting parties can use cryptocurrency as a form of payment related to "smart contracts." To express this point another way, the blockchain records many Bitcoin payments that are made pursuant to smart contracts. A smart contract executes itself when the parties exchange keys, and so there is no question of nonperformance or enforcement as there would be in a standard contract scenario. Following Nick Szabo, Werbach along with Nicholas Cornell has analogized the smart contract to a vending machine.[19] Inserting a coin into the machine unlocks the bargained-for object; once the coin has been inserted, there is no further human action needed in order for the contract to be performed. In the same way, a smart contract is performed upon agreement because the agreement amounts to an exchange of encrypted information. In a more familiar or traditional contract, two parties exchange promises to perform an action: R promises to pay E $50, and E promises to cut R's lawn. Performance occurs in the order specified in the contract—perhaps payment is to be made once the lawn has been cut. But a vending machine purchase (or a smart contract on the blockchain) executes itself. There is no suspense about whether the other party will fail to perform.

Werbach and Cornell articulate what is new about smart contracts, and there is a significant amount of novelty. As enforcement becomes unnecessary, the state recedes into the background, taking with it the data collection and scrutiny related to financial regulation. Freedom of contract is already widely held to be worth protecting, dating back at least to the 1905 *Lochner* case that declared contract-making to be a fundamental right and (earlier) the Contracts Clause of the US Constitution. Smart contracts extend even greater protection to that right—at least in a negative sense by preventing the state from interfering with it.[20] Of course, the absence of the state could create a condition similar to Thomas Hobbes's prepolitical environment, where anarchy predominates and trust has no grounding.[21]

The same features of cryptocurrency that make its use seem questionable—namely, the lack of government backing and the resulting risks—make it an attractive medium of exchange. Government does not support it but does not monitor or regulate it either. Werbach notes that when the municipal government in Buenos Aires forbade credit card payments for Uber services, a platform using Bitcoin allowed Uber to continue operating there.[22] In that case, the blockchain provided a means to avoid government regulation that stood in the way of parties wishing to make a contract. Moreover, transactions are anonymous. There have been illegal uses of the blockchain, some of which did eventually attract government investigation and prosecution.[23] Nonetheless, it has been to date a means to avoid government scrutiny for the most part. Whether cryptocurrency can displace the role of the state in contract transactions altogether is another matter. Illegal smart contracts, like other forms of cybercrime, are prosecuted. In addition, questions about contract terms, or about the fact of agreement, linger and thereby preserve a role for the state in the domain of smart contracts. "Litigation will persist," Werbach predicts, "but it will shift from claims of breach, to claims of restitution."[24]

Cash/Barter Transactions

While cryptocurrency affords its users one kind of anonymity by obscuring the details of transactions from everyone but the transacting parties, paying in cash or bartering can also provide secrecy. Checks, wire transfers, and credit purchases leave a record to be discovered by law enforcement or intelligence agencies, and these capabilities were strengthened post-9/11 by the Patriot Act—specifically, by the "business records" provision.[25] The paper trail left by bank transactions is discoverable by government investigators, but cash payments are harder to track. One possible response to the state's growing ability to follow money routed through banks is to use cash and avoid banking. As with many of the choices discussed in this book, the option of using cash for all purchases and payments is not always a realistic option. Toll roads, for example, have fewer lanes for cash payment as compared to automated toll collection, and some bridges, such as New York's Verrazano, do not accept cash at all. Car rental agencies require a credit card, as do many advance reservations. Often, the best one can do is to limit the number of noncash transactions rather than avoiding them altogether.

A few individuals have taken things one step further by rejecting money itself as a medium of exchange and relying instead on bartering and scavenging as the means to subsist. Utah resident Daniel Suelo decided years ago to forgo the use of money. He lives in a cave and eats only what he can forage from dumpster or desert. He owns virtually nothing and maintains none of the social and economic ties that most people take for granted (street address, phone number, place of work, retirement plan). As his biographer discovered, it took Suelo a long time and much soul-searching to break those ties and to commit to living as he does.[26] He receives no medical or dental care and is left unprotected against illness, weather, and scarcity. He will cut a lawn or provide other services in exchange for a meal. On one hand, Suelo has managed to avoid passing through the nodes at which the rest of us are identified, classified, and sorted: tax filings, earnings record, bank statements, air travel, credit card activity. Fewer records exist that could pinpoint where Suelo has been or what he has done. On the other hand, a person who fails to conform to the behaviors and appearance that are expected and routinely encountered draws attention and even suspicion. Living off the grid but near an established community—even one as relatively free, open, and heterogeneous as

Moab—makes one stand out as a break in the pattern. Despite his lack of a digital footprint, and perhaps even because of it, Suelo stands out.

Learning about Suelo's life, one cannot help but commend and respect him for the stand he has taken. At the same time, one senses the gulf that separates him from other people—even from those who sympathize with him and understand his response to the predicament of living in our consumption-obsessed world with its many entrapments. In his book about Suelo, Mark Sundeen readily admits that he himself could not "quit money"—and that, in fact, some part of him wished at first to turn away from Suelo and from what his scruffy appearance represents: a challenge as well as a rebuke. There is security in the patterns and routines of contemporary life, as well as luxury of not having to worry about how to secure the means to one's survival day to day. The conditions of our world structure and limit some of the larger choices about how to live. Rejecting automobiles, electricity, grocery stores, and money is not impossible, but it is certainly difficult and complicated, and those rejections foreclose other choices.

Surveillance Art

Any discussion of resistance to surveillance must include surveillance art: works in various media, installations, and performance that draw attention to and critically interrogate the proliferation of surveillance technologies as well as public attitudes about those technologies. Some of the artworks amount to countersurveillance in a quite literal sense. "Mr. Security" sets up cameras and microphones in front of state security buildings in order to expose and highlight the work that security agencies are constantly doing. As James Harding notes, this installation is discursively disarming.[27] When asked what they are doing, the artists simply reply, *the same thing you are.* Petr Pavlensky's "Threat: Lubyanka's Burning Door" operates similarly. As thousands of Muscovites pass the old KGB prison each day, they can no longer ignore it once the doorway is on fire.[28]

One of the ends accomplished by surveillance art is to highlight everyday phenomena that we tend to overlook. As a result, we see the extent to which certain moves by the state to accomplish greater control of the population have been normalized. A 2019 exhibit at The Shed in Hudson Yards, New York City, featured a set of works, some of them interactive, that challenged the audience to think critically about technology in general and surveillance in particular. Entitled "Manual Override," the exhibit did not evaluate technology as repressive or liberatory in itself but sought instead to encourage thought and dialogue about it, suggesting that technology-related practices are a "collective mirror." There is no simple solution by which humans can "manually override" technological processes in which they live, but critique is still possible, and critique can lead to more limited, localized intervention and policy change. To that end, a video in the exhibit presented public reactions to predictive policing, in which law enforcement concentrates resources and attention on neighborhoods that have been determined algorithmically to be likely sites of crime. The artist suggested that such policing initiatives amount to "algorithmic violence," constructing people and delimiting their life-chances according to a numerical formula—not to mention raising the stakes and the risks of police-civilian encounters in those "red squares."[29]

In another part of the exhibit, viewers were encouraged to create their own "shadow self" by entering their email address and then watching as a shadow outline of each viewer's body emerged on the screen, complete with demographic data. Alongside the silhouette, other personal details gleaned from the Internet scrolled by: past residences, family members' names, places of employment, and so on . Strikingly, one piece of my own personal data was erroneous: the name of a close relative was inaccurate. Noticing that error, whose source was probably an erroneous transcription by some government agency or private vendor, I realized how a shadow self or data double could mutate in ways that produce real-life consequences. Such errors might obscure a familial connection or create one that does not exist.[30] The Marxist philosopher Louis Althusser claimed that the state's naming, or "hailing," of its subjects "hardly ever miss their man."[31] The subjects respond to the hailing and thereby confirm the identity that it has constructed for them—worker, soldier, and so on. By contrast, data collection and aggregation generate a shadow self that *can* misrecognize a person because the collection and aggregation are not guided by intelligent, critical thought. They are automatic. The importance of surveillance art as a form of resistance lies in its ability to uncover and dramatically convey harms and injustices such as misrecognition. Indeed, the introduction to "Manual Override" suggests that technology is a collective mirror—a mirror that can distort as it reflects.

This chapter has enumerated and described forms of resistance to surveillance technologies. Each of these responses can actually be seen operating in contemporary society. This is not to say that they have neutralized the threats and harms detailed in the preceding chapters. But encryption, countersurveillance, and the like are instances where contemporary individuals exhibit agency in the face of surveillance practices that seem totalizing and irresistible. The fact that resistance is occurring underscores the continuing possibility of agency in the face of the machine. The obstacles and the low rates of participation demarcate the limits of agency.

CHAPTER SIX

Conclusion: Technology, Surveillance, and the Social Contract

> Man is born free, and everywhere is in chains.
>
> —Jean-Jacques Rousseau, *The Social Contract and Later Political Writings*

> Interactive citizens in the war on terror are invited to invest in their own security, both metaphorically in the form of educating themselves to participate in securing the nation, and literally in the form of speculation in the "growth market" of homeland defense.
>
> —Mark Andrejevic, "'Securitainment' in the Post-9/11 Era"

The preceding chapters have explored the trade-offs navigated by contemporary individuals as they decide whether to engage with technological processes and artifacts. A central tension throughout has been between free and voluntary choice, as emphasized in liberal political thought, and the growing structural impediments that we confront as we attempt to employ that free choice in our everyday lives. How much freedom do we actually enjoy as we face decisions concerning technological engagement? This final chapter poses the question, Do we need to rethink our assumptions about the scope of free choice in liberal societies? One such assumption involves social contract theory, which grounds modern liberal-democratic political orders. The notion of the social contract involves the normative claim "that coercive political authority can be legitimized through the notion—either historical or hypothetical—of some kind of pact."[1] First posed by political theorists at the threshold of modernity, the idea of the social contract is used to explain how modern societies with government by consent came into existence. Hobbes suggested that such a state of things never actually existed (the story is merely a heuristic that aids our understanding of constitutional government), while for Locke it was always the default situation—before a popular sovereignty-based government was formed, and from the time one of them dissolved until a new one was enacted.[2] Broadly speaking, the social contract has become a widely shared and accepted part of the "background culture"[3] of modern societies, which is to say that references to the "consent of the governed" can be found abundantly in popular, political, and scholarly discourses.

Since there are numerous versions of the construct of the social contract, it is helpful to specify some of the details, or terms, that form a baseline understanding of it. All versions include the requirement that individuals renounce or forbear from exercising some of the freedom they would otherwise enjoy in a prepolitical state, which would have allowed them to do as they wished. In return for surrendering this freedom, they expect to receive protection from the "coercive political authority"[4] of the state that they have created. The result is a safer and more stable social environment, which is the benefit of the bargain that induces them to form the contract in the first place. In Locke's version, a people constitutes itself by forming a civil society and specifying rules for its operation. Since it is the people themselves who enter into this agreement, the contractual bonds can be dissolved when the governmental order no longer promotes the ends for which it was created. The analogy to a private contact is imperfect, since there are many parties to the social contract (not just two), and they need not be unanimous to dissolve it. Also, the state mediates relations among its citizens who made the contract, but it can through malfeasance lose its authority to rule. Despite these features that distinguish the social contract from a private one, the contract notion is an aid to understanding government by popular consent. And so for more than four centuries, government by consent has been seen as the binding force that holds societies together more enduringly than physical coercion. A ruler can only rule by force alone for so long, but when consent is obtained, a stable and lasting political order becomes possible.[5]

One understanding of the social contract that has been expressed with greater frequency in the post-9/11 era is the "liberty for security" trade-off. Here, the reduced personal liberties allowed in the wake of the 9/11 terror attacks are justified by the need to protect society against new terror threats. Airport passenger screenings, warrantless surveillance, and massive data collection from social media platforms are instances where new or newly strengthened limits have been put in place, and people run up against those limits as they go about the everyday activities of communication, shopping, and travel. There may be grumbling and complaints of inconvenience, but in the end, consent by grudging acceptance—in exchange for expected safety—is the rule. As noted earlier, the government cannot guarantee its ability to make good on the promise of safety, and it even explicitly disclaims its ability to do so at times.[6] Nonetheless, the liberty/security trade-off is simply a contemporary and more specific version of the original social contract: both promise some form of security from physical harm.

Simon Chesterman offers the suggestion that a new social contract is possible in the post-9/11 world of increased, more differentiated, and technologically sophisticated surveillance, and that this contract can offer increases in both liberty *and* security rather than the zero-sum trade-offs of one for the other that we are accustomed to seeing. The key to this reimagining of the social contract is greater government responsibility for privacy protection paired with greater legal constraints on those doing the surveilling.[7] In a similarly hopeful vein, David Lyon has suggested that a "hidden hope" arises from the "participatory turn" to self-surveillance. Lyon believes that individuals living amid (and participating in) ubiquitous surveillance might learn to practice "good gazing" and in turn develop an ethic of care for each other rather than one of suspicion.[8] Each of these prescriptions involves a reimagining of relationships among citizens and between citizens and government. As possible ways forward from the present conditions, they must be seriously considered and assessed for their viability in truly remedying a social contract that no longer provides what it promises. I propose that the best way to undertake such an evaluation is by looking at two levels of social activity where social contract is performed or reproduced: the macro level comprising structural change and governance projects led by state and private-corporate institutions on the one hand, and the micro level of everyday decision making by individual agents on the other.[9]

The Larger Picture: Macro-Level Changes Threatening the Social Contract

At the macro level, the phenomenon of the "corporate contract" poses a threat to the social contract as it is typically understood. This term, used by Sadowski and Pasquale in describing the smart city, was introduced in chapter 4. It refers to the partnerships forged by smart technology firms with city governments, and the authors refer explicitly to its role in replacing the social contract. The implications of this substitution are clear: smart technologies benefit the firms who sell them and the governments that use them as aids to social control. Of course, technocrats would point to the gains in convenience, safety, and so on offered by smart technologies to citizen residents, but the contracts themselves bind and benefit the firm and the state rather than the citizenry. When a city government contracts with a tech firm to incorporate a new functionality into the city's infrastructure, actual legally binding contracts are executed, so that the corporate contract is a literal legal form as well as a metaphorical representation of the state/industry relationship. And these two contracts—social and corporate—stand in a hierarchical rather than a side-by-side relationship to one another. Smart city promotional literature demonstrates the priority of the corporate contract over its social counterpart by omitting mention of citizens as stakeholders, thus making it clear that the social contract has been displaced.

Moreover, there is also a legal argument that is used to support this prioritization of the corporate contract over the social. In the nineteenth century, American courts for a time upheld the contract rights of private parties against government attempts to limit them.[10] Even though this particular constitutional doctrine eventually lost majority support of the Supreme Court, interested parties continued to find ways to elevate private contract rights over the legislatively determined public good. One particularly consequential example from the twentieth century involved Robert Moses, whom I mentioned in the previous chapter. When Moses worked for Governor Alfred E. Smith, he drafted legislation to help promote investment in public works bonds. A clause in the bill that eventually became law declared that the contractual rights of investors could not be displaced by government action. Moses slipped in this statutory language so that construction projects such as bridges and roadways, administered by quasi-governmental entities like the Triborough Authority (which Moses headed), funded the administering entity itself in a long-term and perpetually assured way. This provision accomplished essentially the same thing as the earlier expansive readings of the Contracts Clause did: it elevated the rights of investors over the public good.[11] Thus, the state government could not interfere with the administration of a bond-funded project even after that project was complete. This legal history of the nineteenth and twentieth centuries is recapitulated in the corporate contracts undergirding the smart city today. The neoliberal fantasy of high-tech protection of the citizenry is just that—a chimera obscuring the real beneficiaries of state-corporate partnership.

Another structural change to the social contract can be seen as government enlists "citizen-spies" in "securing homeland defense." This change is structural in the sense that the line separating government from civil society is blurred. The classical understanding of the social contract envisions citizens renouncing freedoms so that the government can protect them. Here, though, citizens assume at least partial responsibility for protecting themselves. Mark Andrejevic (citing Nikolas Rose) likens this move to the creation of "citizen-spies" in the past.[12] In both instances (earlier and contemporary), the baseline duty of the modern liberal state (protection) is delegated to citizens. This development might be seen as a manifestation of neoliberalism because it privatizes a state function, but there is more to say. People policing each other and themselves makes for a more effective society of control in the same way that penal discipline, for Michel Foucault, was strengthened by the move from punishing the body to punishing the soul, in his terms. Foucault famously charted the increasing sophistication and effectiveness of the political technologies of power that ushered in a phase of governmentality wherein subjects police themselves.[13] As prisoners and other members of society internalize behavioral norms conducive to state control, those norms become stronger and harder to challenge. The same is true of homeland defense. Andrejevic makes this link in his description of what he calls "securitainment": homeland defense learned through popular cultural forms such as security-related reality TV and then practiced as citizens encounter each other in everyday interactions. In the United States, police, transportation authorities, and public parks officials encourage citizen participation in detecting terror plots by running public service announcements enjoining them to "say something" if they "see something." There is a potential, in this arrangement of things, for an even greater level of distrust among citizens than the traditional civil society envisioned by Locke would produce. There, we ceded the role of executor of natural law (which we all have in the prepolitical state) over to the government once the social contract was in place. We might eye each other with suspicion, but disputes would be referred to the government for peaceful resolution. Once citizens are charged with detecting and referring their neighbors' suspicious behavior to the investigating state authority, however, the climate changes and distrust increases.

Moreover, the popular culture dimension of this inquisitorial/investigatory project makes matters worse as it brings imagination and affect into play in damaging ways. As Andrejevic shows, reality TV draws upon public fear, anger, insecurity, and frustration to mobilize the public into anti-terror investigators. The figure of the dangerous foreigner infiltrating national borders is as old as the idea of the nation-state itself, and it is extremely serviceable as a result.[14] Now, in the world of new surveillance, that figure can be seen daily by those who wish to look, thanks to reality TV and Internet technology. Andrejevic describes an Australian reality TV program in which border security threats are detected and neutralized. Fantasy merges with reality and helps to *shape* reality, thus enacting the recursive relationship between surveillance imaginaries and concrete social practices depicted by surveillance scholars.[15] A state-funded program in Texas takes reality TV a step further by allowing Internet viewers worldwide to monitor specific points along the Texas/Mexico border and to report people crossing into the US there.[16] This is certainly a postmodern moment: viewers on their couches (or in a pub in Australia, as Andrejevic reports)[17] engage in a form of play not unlike a video game except for the fact that the consequences of the game are life-altering for the people on screen. And this is not simply a matter of citizens policing their own borders, since viewers around the world, citizens of other countries, are participating as well. It hardly needs mentioning that the objects of the surveillance game are more often than not racialized Others. Thus, a transnational public is formed and united in opposition to an exoticized figure that serves as a repository for that public's resentments and anxieties.[18]

The social contract is also threatened, at the macro level, by the possibility of what political philosopher Isaiah Berlin called a "liberal-minded despot."[19] Personal liberty and self-government are presumed to go hand in hand and to require each other. In one sense, this is what the social contract promises. While subjects surrender a certain freedom to do as they wish when they form the social contract, they do so in order to secure for themselves a more permanent liberty, a space in which to pursue their visions of the good life without threat from the government or from each other. This association between liberty and democracy often goes unquestioned, as if it is self-evident. A half-century ago, Berlin urged readers to look more closely at the liberty/democracy relationship. He noted that in principle liberty could coexist with a despotic regime. Berlin thought it "perfectly conceivable that a liberal-minded despot would allow his subjects a large measure of personal freedom. The despot who leaves his subjects a wide area of liberty may be unjust, or encourage the wildest inequalities, care little for order, or virtue."[20]

Today's neoliberal regimes, such as the United States, fit the bill of the liberal-minded despot strikingly well. Of course, the United States is in formal terms a constitutional democracy, yet neoliberal state-corporate and military-industrial partnerships lead it to function in "unjust ways" and to "encourage the wildest inequalities," thus matching Berlin's description of despotism. Millions of Americans are incarcerated, and tens of millions live in poverty. At the same time, a "large measure of personal freedom" is enjoyed by many people. In the role of consumer, Americans may choose from an array of sophisticated products. Even people living in poverty can to some extent exercise such highly specific and preference-based choices. Television streaming, soft drinks, fast food, cell phone capabilities, and designer clothing offer a certain kind of freedom, a freedom to make personal choice from among many products on offer. Even the decision to spend one's leisure time watching reality TV, or surveilling border-crossers and phoning them in, is an enjoyment of a certain form of free choice. The forces of consumerism have reshaped the social contract into a tacit agreement that reads as follows: *citizens will consent to (or at least refrain from challenging) a central state that rules in unjust ways as long as certain areas of personal freedom (especially consumer choice) remain available*. Once again, this agreement is prevalent but not absolute. Citizen organizing, protest, and other forms of activism are still possible, and they are practiced. On the whole, though, many people opt for what Pramod Nayar in chapter 2 called "consumer citizenship," and we see people living out its terms all around us. I will return to a normative evaluation of this trade-off shortly; for now, I want to highlight it as one of the changes to the idea of the social contract.

The macro-level threats described above could be mitigated if individuals could opt out. Chesterman explores this possibility—specifically, that citizens may find ways to escape personal data collection by private corporations and communication monitoring by government. I have also referenced in the preceding chapter the same kind of potential avoidance in connection with the notion of irresistible surveillance. Email encryption, or forgoing email communication altogether, are steps people could potentially take to retain control of personal data and minimize their own data footprints. I include this topic here under macro-level changes to the social contract because questioning whether it is possible to opt out helps to elucidate the structure of the social contract as it now exists. In traditional social contract analysis the population is bound by the contract even if they did not agree to it, or sign onto it, as individuals. After all, each person can seek to benefit from the contract on a given day by using the courts, calling the police, or navigating public infrastructure. Chesterman is questioning whether the new social contract, in which liberties are curtailed in order to provide security from a terror attack, is mandatory or inescapable. This is a difficult question to answer in the abstract. Nonetheless, a few observations are possible.

First, national security frequently forecloses the possibilities for self-protection because it carries an overwhelming force that is backed by certain legal doctrines as well as historical precedent. In the Japanese internment cases of the early 1940s, the Supreme Court deferred wholesale its institutional power to resolve disputes involving questions of constitutional liberties, preferring instead to leave the matter to military discretion.[21] This deference has been seen again in the some of the cases arising post-9/11 out of the Patriot Act and other anti-terror laws.[22] Deference to government can also be seen in the use of the *state secrets privilege*, a seventy-year-old evidentiary doctrine created by the Supreme Court early in the Cold War, which allows government to refuse requests—and often terminate litigation entirely—when the government raises national security concerns about potential evidence.[23] "State secrets" cases show perhaps more clearly than other disputes the way in which "national security" can be used to override individual privacy concerns, making it impossible to opt out of the data collection and other surveillance that the government undertakes against individuals and groups under the heading of national security. The "new" social contract involving a trade-off of liberty for security is inescapable for citizens because the claim of national security can sweep aside complaints and efforts to stop rights violations. "State secrets" has been used to block lawsuits challenging spying on children, suing for torture, and obtaining medical treatment for government-caused illnesses. Currently, the ongoing battle between the federal government and Apple over unlocking iPhones implicated in criminal investigations raises the national security question. The conflict pits imperatives of corporate profit (in turn dependent on brand trust) against anti-terror investigations. So far, courts have had trouble resolving those cases.

Another factor working against individuals who try to escape ubiquitous technological engagement under the new social contract is the government's ever-more-aggressive use of the technology itself. The Edward Snowden disclosures revealed the NSA's almost obsessive focus on collecting all possible data. The documents he leaked show how the agency was determined to find new ways to capture what it didn't have before, to "collect it all." This contest between a government seeking to obtain all existing personal data and a public trying to find ways to retain control of their data plays out endlessly, like the nuclear arms race.[24]

A Granular View: Agential Choice under the Current Version of the Social Contract

The terms of the social contract structure a political order and establish civil society. None of us actually executed such a contract at a discrete moment in time, but we live out its terms by participating in government, following laws, and not rebelling. We also enact a version of the social contract by striking smaller bargains that are *nested* inside of the larger one. In a sense, every engagement with a technological artifact is a replication of the social contract—at the individual level. As I mentioned earlier, airline passengers every day decide whether to submit to X-ray and electronic screening (or miss the flight), and this bargain is nested inside the liberty/security trade-off enacted by homeland security laws, which is in turn nested inside the social contact itself. When a driver decides to sign up for EZ Pass, and when an airline traveler signs up for PreCheck status, they are also enacting a nested bargain since the government makes these programs possible in the first place and authorizes them through various laws and regulations. Some interactions are more difficult to characterize as nested bargains, either because individuals do not know they are happening (and therefore cannot accept or refuse) or because they know what is being surrendered but are powerless to prevent it. Even in the examples given, the choice to participate can be illusory to differing extents, as refusal may not be a realistic option. Thus, levels of freedom and unfreedom are reproduced by nested bargains people make every day. By submitting to airport searches, passengers reproduce a form of the liberty/security trade-off that reinforces the government's ability to search and to know more about citizens without a guarantee of increased safety. When social media users continue to post content even though they know that the content may be turned over to security agencies, they are reproducing, through a nested bargain, an arrangement whereby corporations collaborate with government to diminish privacy. These individual micro-level choices contribute to a change in the structure of state/citizen power relations.

Social media use facilitates data collection but also generates responses at the level of individual choice: individuals might choose to use communications technology—particularly social media platforms—to *present the self strategically*.[25] A preoccupation arises wherein people focus inordinately on representations of the self mediated by communications platforms. This preoccupation cuts across contexts from professional to entertainment to personal. Photos, concise messages, reactions, ratings, endorsements, and references to other sources all contribute to what Pasquale calls "grooming" the self[26] and Lyon describes as creating a "strategic self."[27] Scholars studying new surveillance emphasize *participation* by subjects in surveilling themselves and others, and they also note that some of this behavior is strategic. Individuals want to project a certain representation of the self that is useful or helpful to them in some way, whether as means to a job or a date or simply as a version of the way they wish to be seen. Individuals enjoy agency in this process of creating an online self: they can tell stories about themselves and present flattering, selectively processed, mundane, and even incriminating images for other viewers to consume. In a sense, the possibilities for self-disclosure provided by social media allow us an unprecedented degree of control over defining ourselves and shaping how others see us. The extent to which people attempt to exercise this agency can be confirmed if one does not mind appearing nosy. Go to a public place such as a coffee shop or train station and note how many people are using their cell phones (everyone). Then see how many are communicating via some form of social media (as opposed to streaming content or talking in real time). Many of the people you see will be engaged in some form of social media–related disclosure or commenting on what someone else has shared. The process is continuous and can be seen happening everywhere there are people.

While individuals groom their online selves through deliberate and volitional choice, we know that the subjectivities of social media users are shaped by the platform owners and controllers. When the infamous Facebook emotion experiment mentioned earlier was made public (the fact of its existence, that is, not the specific methodology), the truth that the experiment manipulated users became undeniable fact. As Facebook customers decided what to post or comment on, their responses were being studied and recorded. And even before the public knew about that experiment, we knew that online shopping data was being used to market more precisely to customers. All of this suggests that there is a limit to the autonomy that social media and Internet users actually enjoy. There is a complex relationship between online freedom and online conditioning. Internet engagement is a means of interacting, a space like the public sphere of early modernity. The Internet creates a (virtual) space where individuals can pursue chosen ends—just as the social contract is supposed to do. It is at once a space that allows for the practice of liberty (specifically, the liberty to develop and disclose the self) and also a means to be used, as we have seen, for political organizing and commerce as well as social networking. It is undeniable that this wide range of strategic behavior is possible and that individuals do engage in it. But like other consumer behavior, the seemingly free choices of social media users are limited by the contours of the platform itself. My "profile" differs from the next person's in our specific choices of content, but all of the options for how content is displayed are set for us in advance by the platform owner. Anyone who uses social media to shape a presentation of the self, to develop a reputation or evaluate the reputations of others, is doing so according to a pre-given set of rules and limits.

All of the trends toward greater surveillance and consumer conditioning noted by commentators on digital privacy have taken shape over time and have accelerated with the advent of Internet communications technologies. While commentators all notice these trends, some see hope and possibility for positive change. Lyon suggests that opportunities for questioning, negotiating, and even subversion may open up. Such participation offers chances for making rights claims that can contribute to the overall shaping of the Internet or for contributing to the mitigation of unnecessary security procedures—small but significant acts that could stimulate shared practices.[28]

Lyon offers this hope in opposition to the defeatist and fatalistic discourses about the inevitable loss of personal privacy in a totalizing society of control. And it is undeniably true that contestation is possible even in a world of pervasive and complex surveillance infrastructures. Recall "Mr. Security" from the previous chapter, a performance art piece in which cameras are trained on state security posts. By pointing out these "cracks" in the social architecture, Lyon is echoing the observations expressed in deconstructionist criticism and postmodern theory more generally to the effect that meaning can never be fully and permanently fixed through language, and that no attempt to fully enclose social reality can succeed.[29] Some remnant, surplus, or kernel always escapes. To deny this possibility would be to ignore the many instances in human history when grassroots movements impacted national politics and even led to regime change. Such moments were what Hannah Arendt may well have had in mind when she said that humans are capable of the "infinitely improbable."[30] And in any event, it is difficult to live without hope for positive change.

Without denying that there is a reason for hope, as well as a human need for it, a question remains to be asked from the other end of things: why haven't we seen mass opposition to government surveillance or consumer surveillance? This question can be broken down into smaller and more specific ones:

- Why do consumers so casually give away privacy for short-term gains in convenience and efficiency?
- Why did Edward Snowden's disclosures of massive government data collection provoke so little public outrage (more people were angry with him for disclosing than with the government for its actions)?
- Why has the Internet come to function more as a means to mundane communication and consumption than as a liberatory tool?

In other words, while it is important to ask what possibilities for resistance exist, it is equally important to ask why power arrangements are already constellated as they are. If there are viable avenues of escape, why aren't people escaping? These questions are extremely difficult to answer in a satisfactory way. In fact, they resemble some of the most basic inquiries of political and social theory. Considering more general questions about the workings of power, Michael Rosen says (recalling Wilhelm Reich) that a modern theory of power needs to explain why hungry people often *don't* steal food, why oppressed workers often *don't* strike, when it would be clearly in their interests to do so.[31] These foundational questions are useful to ask here not because there is a simple answer, but because asking them helps to get at the more powerful forces at work in blunting the edges of resentment and moving people toward acceptance instead of resistance.

One factor that Lyon himself notes, drawing upon Deleuze and Guattari, is the development of "a culture of desiring machines."[32] Individuals as consumers function according to their desires, and subject position of consumer becomes most important, infiltrating other realms of our lives. Here are some of the many questions whose answer has become: "Consume!"

- How can I show my family I love them?
- How can I gain status in my community?
- What should I do during my vacation?
- What's the best way to fight terrorism? (This question was posed and answered by President Bush in 2001.[33])
- Wal-Mart is "critical infrastructure" because it protects our right to do what?[34]

The machine is often depicted as an engine that generates consumption, and one of its effects is to produce a society of consumers. Critiques of the consumer society are not difficult to find, as they come from many quarters simultaneously. Long ago now, Herbert Marcuse described, in *One-Dimensional Man*, the process by which all aspects of modern life, including leisure, have become marketing opportunities. Habermas, too, writes of "the soft compulsion of constant consumption training."[35] In his important work *Democracy's Discontent*, Michael Sandel sounds a similar anticonsumerist complaint in a civic-republican register. Sandel means to criticize proceduralist liberalism for its emphasis on rights and its neglect of duties. By remaining neutral on the relative value of competing conceptions of the good life, liberalism generated political institutions that were judged according to their capability to allow individuals to pursue their chosen conceptions of the good. As Sandel puts it, "The emerging consumer society held out an alternative, privatized vision of freedom less demanding than the republican vision of freedom as self-rule."[36] Sandel fears that liberalism has led American democracy into crisis by defining freedom so narrowly and encouraging the pursuit of selfish ends. He sees a way out in the cultivation of institutions that promote civic virtue. By refusing to acknowledge virtue in a substantive sense, liberalism left a moral vacuum that was quickly filled by the market, which established consumption as the primary concern for citizens.

Commenting on Sandel's work, Jean Elshtain and Christopher Beem have explicitly linked the problem of consumerism with bargaining. "We have made a political bargain with the state," they write. "We have paid for the establishment of an atomizing conception of freedom by sacrificing a measure of civic autonomy."[37] They elaborate on this bargain as follows: "So long as their most basic and even base desires are fulfilled, the citizenry would one day permit the state to quietly dismantle political liberty."[38]

This is a surprising and unfamiliar form of the social contract. We are accustomed to a more salutary conception of the bargain, struck by individuals in the state of nature, by which modern constitutional government came to be. As typically articulated, the social contract would *encourage* rather than preclude civic involvement. Individuals would have a stake in seeing that the government is properly limited in the scope of its power and that its functioning aligns with the popular will. Here, though, the contract has devolved into something else. It leads to "the replacement of citizenship with a state-sanctioned, consumerist-driven, perpetual childhood."[39]

In advocating civic republicanism as the antidote to this false freedom, Sandel makes clear that in his view, the turn toward consumerism was neither inevitable nor incurable. In the long-running quarrel between liberalism and republicanism, he sides with the civic republicans and blames liberalism, with its emphasis on individual rights, for the "discontent" many people feel with the current state of American democracy.

I am less interested, here, in continuing the liberalism/republicanism debate. Rather, I have cited critiques associated with Sandel in order to outline the role of consumer behavior (and of desire more generally) in reproducing some form of consent to a surveillance society. The force of consumer desire, and the way desire has been mobilized by platform owners and other profit-seeking firms, is part of the reason why people make bad bargains or fail to object to arrangements that are no bargain at all. Consumer desire must be reckoned with if we are to rearrange the constellations of power that constitute the surveillance society. The civic republican critique gives less attention to the role of the *state* in producing conditions of diminished privacy, but this is because it leans harder on civic duty and de-emphasizes assertion of rights. But the smart city initiatives and the militarization of policing effect an increase in state power and a diminution of citizen control over government, and they too unquestionably play a role in the failure to oppose the surveillance society. These two state activities also represent a convergence of state and corporate interests, of profit and power.

Normative Judgments about Social Bargains

When we think about bargaining, we tend to envision cost-benefit analysis. This is hardly surprising in light of the prevalence of rational choice theory and microeconomic modeling. Cost-benefit analysis is often used as a paradigm for policy making (especially in environmental policy); this was true during the Reagan years in particular, but the approach remains in use today.[40] Specifically, cost-benefit analysis in the regulatory context requires that any policy action to be undertaken must first be proven warranted by producing a decrease in future monetary costs to government (i.e., a benefit) that exceeds the cost of its implementation.[41]

Critics of the cost-benefit analysis approach to policy making point out its flaws. Some costs (e.g., species extinction) cannot be easily monetized and therefore fall outside the calculus. Future catastrophe that is unlikely but still possible also escapes the cost-benefit formula. Natural features (e.g., barrier islands that provide flood control) go unacknowledged and face destruction as a result.[42] In his recommendations for future disaster prevention policy, Robert Verchick makes a convincing case for other frameworks: precautionary analysis and scenario planning, for instance.[43] He also demonstrates the relational aspects of property rights and land use. People hold and use land in relation to one another and in the context of their personal relationships, a fact that eludes a calculus based solely on cost comparison.[44] In a similar vein, Joseph Singer evaluates welfare maximization and fairness as guides to social policy, and he reveals that welfare maximization (an approach that seeks to maximize the satisfaction of people's stated preferences) similarly sidesteps questions of fairness rooted in a system of values. This is another way of saying that a simple cost-benefit analysis is inadequate to the task of deciding how best to distribute public goods. The point here is that even at the policy level, where budgetary planning requires consideration of monetary expense, cost-benefit analysis has its limits.

Moving from the level of government policy making to individual decision making, we see that the complexities of cost-benefit analysis become even greater. Monetized costs are certainly part of the calculation as to whether a given bargain is worth making, but they are only a part of the picture. In chapter 2 the discussion of convenience and efficiency referenced possible savings of effort or money. Customer loyalty cards promise savings; online surveys are rewarded with coupons for free food. These are specific instances where a monetary calculus fits. But the challenges to a thorough public-level decision-making analysis highlighted by Verchick and Singer are exacerbated when individuals attempt that analysis. Most obviously, governments have superior information-gathering capabilities as compared to individuals. There is a greater chance that information will be hidden from the individual bargainer. At times this is the result of deliberately secretive behavior, as when Target utilized secret algorithms to market baby products to pregnant women. Also, however, some information is uncertain in itself, even if no one is trying to hide it. In chapter 3 the vagueness of harm was shown to disadvantage consumers who surrender personal information online. Before the fact of disclosure, the uncertainty of future harm prevents people from seeing what they risk by providing information, and after the fact the inability to prove a concrete harm in the present moment can preclude damage recovery. Some catastrophic harms are too traumatic to contemplate, and if they have never been seen before they are even more difficult to visualize. This ambiguity about the future affects individuals' ability to calculate costs and benefits. Finally, individuals do not simply decide about a bargain based on rational factors but through the influence of affect as well. Sellers of goods and services mobilize affect and shape desires to create consumer subjects who will buy from them.

Throughout this book I have tried to show how individual subjects navigate free choice and structural constraints. An agent-centered perspective such as the one I have offered here effectively conveys the phenomena associated with attempts to exercise choice in a world of pervasive state and corporate surveillance. In an important sense, it is necessary to take individuals at their word when it comes to understanding the bargains they make. This is the imperative of a political order in which people are given space to pursue projects of their own choosing, and one in which the government remains agnostic about the relative value of the projects people choose.

It is easy to see how cost-benefit analysis became established in a political community like the United States that emphasizes individual choice, just as it is easy to see how, in the nineteenth century, freedom of contract took hold here as a principle to be protected.[45] The notion that such autonomy ought to be inviolate has deep roots in the political culture. Thus, when a person chooses to live in a gated community, purchase a technologically sophisticated automobile, or divulge personal information online, we are loath to criticize the terms of the bargain; we leave that to the bargainer. Freedom of contract means that a party may contract for anything that is not illegal; a "peppercorn on New Year's Day" is acceptable consideration[46] if that is what a contracting party expected to receive. Under the freedom of contract doctrine, then, normative judgment of a bargain is difficult, and even suspect. Earlier in this book I suggested that hidden information and unequal bargaining power might give reason to criticize a particular bargain as a bad bargain, or no bargain at all. Those cases resemble the fraud doctrine and consumer protection statutes, which contract law does recognize: fraud by one party can void a contract, and vastly unequal bargaining power can render a contract unconscionable and there unenforceable.[47] Thus, fraud and unequal bargaining power are already part of the law of contracts, and it takes little imagination to transpose them to the social bargain context we are considering here.

There is, however, a difference that arises in social bargains that extends beyond the way we evaluate contracts today, and it leads to a normative critique of social bargains. Similarly to the way Verchick urges planners to look beyond cost-benefit analysis in environmental policy, bargains involving technology and its social effects raise normative questions—about values, about human relationships, and about protecting dignity. The bargains discussed in this book all shape social values. Joseph Singer criticizes the primacy of welfare economics, with its emphasis on cost-benefit analysis and maximizing satisfaction of preferences, on policy making. Reacting to other scholars who favor welfare economics, Singer shows that we do, and we must, ask questions about values when making social policy. He bases this claim on several assumptions. First, people's preferences (what they say they want) are bound up with their values. I might prefer to have my grown child live near me because I think family ties are important. Second, there might be certain outcomes that do not harm me directly but that I would nonetheless be unable to accept. Thus, my values would not be subject to sale even if I preferred to have money for, say, a boat.[48] The upshot for Singer is that societies need to address values because in doing so they ask and determine what kind of society they want to be. For our purposes here, Singer's argument is helpful in understanding why a bargain struck by an individual who engages with technology might be open to criticism for the values that it implicitly promotes. Bargains that help to structure a society (social media use, surrender of information, privatized surveillance, state-corporate surveillance partnerships) are open to question with regard to the ways they help to shape social and political values.

Individual bargains also produce externalized effects. An "externality" in economic theory is a cost borne by someone outside of (external to) a contract. If I sell my factory to a new owner, the pollution it produces will make nearby residents fall ill. Those neighbors will suffer a negative externality. The externalized effects of social bargains work something like this but also bring social effects that are not always physical or immediate but are real nonetheless. Don Mitchell has written about "SUV citizenship," a phenomenon whereby individuals isolate themselves from social interaction, generating more atomized social spaces. Mitchell claims that SUV citizenship was stimulated by Supreme Court decisions creating "bubbles" or "buffers" around people so that others must keep a distance.[49] Consumer-citizens enact their preference for insulated and isolated lives in which they need not hear unwanted opinions (buffer zones), they need not interact with the wider public (gated communities), and they travel higher and safer on the roads in sport-utility vehicles. As Campbell succinctly puts it, "SUVs achieve their relative safety by externalizing danger."[50] The SUV owner is safer than drivers in smaller cars, and the SUV itself poses a danger to other vehicles by its size and weight; a collision with an SUV is more likely to inflict serious injury or death. The choice of greater safety comes at a cost of greater risk to others—strangers who have no say in what other people drive.

This prioritization of the SUV is simply a further elaboration of "automobility": an approach to city and infrastructural planning designed around automobile users.[51] As Campbell explains, "Embodying a functionalist view of the city as an organized machine, American urban planners from the 1920s on relied on a system of zoning controls that separated uses and imposed homogenous criteria on specified areas."[52]

Automobility imposes numerous negative externalities on others: pollution, inferior public transit, pedestrian deaths—and these are only the effects within the community. Campbell shows that automobility requires foreign oil, which in turn leads Americans to construct a foreign enemy who stands in the way of the needed resource. Foreign policy is shaped by this social imaginary, even to the point of war.

A final normative ground for criticizing bargains that sacrifice one's own privacy has to do with the obligation to protect the dignity of the self. Philosopher Anita Allen offers an argument against surrendering privacy through the kinds of bargains I have been discussing here.[53] After noting that "a new, technophilic generation appears to have made disclosure the default rule of everyday life, and it cannot imagine things any other way," Allen states a case for privacy based in duty to the self.[54] She is concerned not merely with the value of privacy in a society (that is, what beneficial effects it might generate) but with "the ascription of ethical responsibility: in addition to any moral obligation to protect others' information privacy, do individuals also have a moral obligation to protect their own information privacy?"[55] Allen thinks we *do* have such an obligation. It is, of course, more clear that we have an ethical duty to safeguard the privacy of *others* in view of the harms that failure to do so would cause them. And it is equally clear that individuals have second-order duties to protect the self in fulfillment of first-order duties to others: a lifeguard must stay fit (second order) so that he or she can protect swimmers' safety (first order).[56] As to the self, furthermore, it is often *prudent* to protect one's information privacy—in order to prevent loss of money, harm to reputation, and so on.

But Allen goes beyond each of these claims and states a first-order duty to the self that exists "over and above prudent self-interest."[57] In her view, it is sometimes wrong to cede our information privacy to the state or to others even if we appear to be gaining something by the exchange. The kind of harm that we suffer in such a bargain is a "dignitary harm," according to Allen. She suggests that "privacy is a requirement of our freedom, our dignity, and our good character," the same as the duty not to lie.[58] It can be a first-order duty and not simply a duty regarding others. We *ought* to value our own privacy just as we ought to value freedom and just as we ought to value keeping our word. To disregard these aspects of being human is to diminish our own humanity. "We are agents and beneficiaries of our own flourishing," she says.[59] To help us visualize what is at stake and why it is important to fulfill this duty, Allen offers the example of a cancer patient who broadcasts details of diagnosis and treatment to coworkers and even strangers.[60] The subject in this scenario inflicts harms on the self, demeaning it by indiscriminately publicizing private facts and presenting the self in a disrespectful and undignified manner. This is not the place to develop or interrogate Allen's argument fully, but it helps us to see how we might not fully grasp what we are surrendering in some of our interactions with technological artifacts. Loss of dignity is a harm that individuals might or might not perceive, and it is at least worth asking whether a person is aware of that loss, and how a loss of one's dignity might be viewed by others. Health privacy is closely connected with dignity, and Congress has enacted a law (HIPAA) that requires express consent to release sensitive health records. This consent provision protects individuals from disclosure against their will, but it also requires deliberation and awareness—whether or not the individual would have manifested them in absence of the law.[61] While HIPAA obviously creates legal duties on the part of health-care providers to protect the privacy of others, its waiver provisions can also be seen as the regulatory reinforcement of a self-regarding privacy obligation. This self-regarding privacy obligation—rooted in the importance of preventing dignitary harms to the self—is implicated whenever people trade away privacy for convenience, efficiency, or other gains with a resulting loss of dignity such as that resulting from the oversharing of medical photos referenced above.

The three normative grounds I have just described—social values effects, externalities, and the obligation to protect one's own dignity—allow for criticism of the technological bargains individuals make, even if those bargains were conscious and deliberate and uncoerced. These normative critiques stand alongside the "bad bargains" described earlier, in which costs are hidden/poorly understood, or where the inability to refuse makes the bargain no bargain at all. My hope is that this framework for examining encounters with technological artifacts and processes will generate useful conversations about everyday interactions with surveillance, communication, and consumer technologies. The philosopher Slavoj Žižek reminds us that "Even if people do not take things seriously, even if they keep an ironic distance, they are still doing them."[62] People interact with technologies and invite surveillance with overwhelming frequency in contemporary social life. This book is an invitation to take those decisions more seriously.

NOTES

Chapter 1. Introduction: Irresistible Bargains

1. James Brooks, “Cyborgs at Work: Employees Getting Implanted with Microchips,” https://phys.org/news/2017-04-cyborgs-employees-implanted-microchips.html (accessed April 3, 2017).
2. Michel Foucault, “The Body of the Condemned,” in *The Foucault Reader,* ed. Paul Rabinow (New York: Pantheon, 1974), 173.
3. Donna Haraway, *Simians, Cyborgs, and Women: The Reinvention of Nature* (New York: Routledge, 1991), 149.
4. Thomas Carlyle, “The Age of Machinery 1829,” http://sourcebooks.fordham.edu/mod/carlyle-times.asp (accessed June 21, 2017).
5. Adam Smith, *The Wealth of Nations* (1776), Book One, Part One, https://www.marxists.org/reference/archive/smith-adam/works/wealth-of-nations/book01/ch01.htm (accessed January 7, 2020).
6. While the “rational actor” is a central part of modern economic theory and its offshoots, there is a danger of oversimplifying how economic choice actually happens. Mark Sagoff has shown that it is impossible to delineate fully the range of costs and benefits flowing from a given choice. See Mark Sagoff, *The Economy of the Earth* (New York: Oxford University Press, 1988). For one thing, a full cost/benefit calculation would require perfect information, which actors never have.
7. David Lyon, *The Culture of Surveillance* (Cambridge, MA: Polity, 2018), 57.
8. Jurgen Habermas, “Discourse Ethics: Notes on a Program of Philosophical Justification,” in *Moral Consciousness and Communicative Action,* trans. Christian Lenhart and Shierry Weber Nicholson (Cambridge, MA: MIT Press, 1990), 43–115.
9. Gary Marx, “Soft Surveillance: The Growth of Mandatory Volunteerism in Collecting Personal Information—‘Hey Buddy Can You Spare a DNA?,’” in *Surveillance and Security,* ed. Torin Monahan (New York: Routledge, 2006), 37–56.
10. Sarah Igo, *The Known Citizen: A History of Privacy in Modern America* (Cambridge, MA: Harvard University Press, 2018).
11. See for example N.J.A.C. 10:90-2.2(a)(1). On the compulsory extraction of personal data from welfare applicants and recipients, see Virginia Eubanks, “Technologies of Citizenship: Surveillance and Political Learning in the Welfare System,” in Monahan, *Surveillance and Security,* 89–107. Eubanks reports recipients’ worries about improper sharing of their personal information, supported by actual narratives in which such disclosure actually occurs. In addition to this very realistic concern, recipients face the frustration of being unable to seek recourse, or even to prove that such improper disclosure occurred.
12. The contextual differences between vulnerable and less vulnerable actors in bargaining scenarios have been noted by others. Nancy Campbell, for instance,

shows how the growth of workplace drug testing from the 1980s onward has continued to fall more heavily on lower-income and lower-wage workers, so that it arises more often in their lives and is more likely to be obligatory for them. "Consenting" to be tested is not really a choice because it is predicated on keeping a job that they cannot afford to lose. As she puts it, the problem is "the imposition of new bodily surveillance regimes in contexts where there is already so little regard for civil rights or social justice that being drug tested without knowledge or consent can become a matter of course." Nancy Campbell, "Everyday Insecurities: The Microbehavioral Politics of Intrusive Surveillance," in Monahan, *Surveillance and Security,* 57–75.

13. Torin Monahan, *Surveillance in the Time of Insecurity* (New Brunswick, NJ: Rutgers University Press, 2010), 23.
14. Marx, "Soft Surveillance," 55n15.
15. Charles Duhigg, "How Companies Learn Your Secrets (Psst, You in Aisle 5)," *New York Times Magazine,* February 16, 2012, MM30.
16. Cory Doctorow, "Privacy, Public Health, and the Moral Hazard of Surveillance," *Guardian,* May 21, 2013.
17. Jay Moye, "Reinventing the Fountain Experience: Coca Cola Freestyle Crosses 40,000 Installs, Continues to Innovate," http://www.coca-colacompany.com/stories/reinventing-the-fountain-experience-coca-cola-freestyle-crosses-40-000-installs-continues-to-innovate (accessed April 19, 2017).
18. *Lawrence v. Texas,* 539 U.S. 558 (2003).
19. Frank Pasquale, "From Territorial to Functional Sovereignty: The Case of Amazon," *Law and Political Economy* (blog), https://lpeblog.org/ (accessed October 31, 2018).
20. Pasquale, "From Territorial to Functional Sovereignty."
21. See, for example, Mary Madden, "Public Perceptions of Privacy and Security in the Post-Snowden Era," http://www.pewinternet.org/2014/11/12/public-privacy-perceptions/ (accessed July 2, 2018).
22. James Harding, *Performance, Transparency, and the Cultures of Surveillance* (Ann Arbor: University of Michigan Press, 2018), 4.
23. Harding, *Performance, Transparency, and the Cultures of Surveillance,* 4.
24. Harding (234) notes the performative character of surveillance itself. Surveillance is performed, for example, by informants who play a role, assume a false identity, and draw their surveillance targets into that role-playing. In response to myriad forms of surveillance performances by the government, some activists have created performative art projects that highlight surveillance practices. In particular, Harding notes "Mr. Security," an installation placed outside certain government buildings to record the work of guards and cameras and then play back the recordings in front of those same buildings. This action serves to remind the public of the taken-for-granted surveillance activities occurring all around them, all the time. A degree of agency is realized by the artist, though of course the effect on power relations between the individual and the state is not immediately known.
25. Simone Weil, *Oppression and Liberty* (New York: Routledge, 2001), 102.
26. Gilles Deleuze and Felix Guattari, *Anti-Oedipus* (London: Continuum, 2004), 1.
27. Or "The Machine" in the TV series *Person of Interest,* as described by Lyon, *The Culture of Surveillance,* 47.

28. Nikolas Rose, "Government and Control," *British Journal of Criminology* 40 (2000): 321–339, at 324.
29. Pramod Nayar, *Citizenship and Identity in the Age of Surveillance* (Delhi: Cambridge University Press, 2015), 150.
30. David M. Wood, "Vanishing Surveillance: Securityscapes and Ambient Government," March 20, 2014, http://pactac.net/2014/03/vanishing-surveillance-securityscapes-and-ambient-government/.
31. Jathan Sadowski and Frank Pasquale, "The Spectrum of Control: A Social Theory of the Smart City," *First Monday* 20, no. 7 (2015): 1–22, at 1.
32. Sadowski and Pasquale, "The Spectrum of Control," 1.
33. Daniel Smith and John Protevi, "Gilles Deleuze," *Stanford Encyclopedia of Philosophy*, https://plato.stanford.edu/entries/deleuze/ (accessed January 6, 2020).

Chapter 2. Technologies of Convenience and Efficiency

1. Jeffrey Jonas, "The Surveillance Society and the Transparent You," in *Privacy in the Modern Age: The Search for Solutions*, ed. Marc Rotenberg, Julia Horwitz, and Jeramie Scott (New York: New Press, 2015), 93.
2. Consumers are not generally free to negotiate terms here. In contract law, the term "contract of adhesion," or "standard form contract" is used to describe such an agreement, in which one party offers a standard contract to the other party, who can then accept those terms wholesale or reject the offer in its entirety; there is no opportunity to bargain over terms or change anything in the written contract. Contracts of adhesion indicate an imbalance in bargaining power between the more powerful seller (e.g., big-box store selling an appliance) and the more vulnerable buyer (individual consumer). When contracting for goods and services via the Internet, consumers face an even stronger form of this phenomenon. If I want to go on in the transaction (or install an app I have purchased), I must "click" to agree. If I don't click on the "accept" box, I cannot even continue the process; I cannot even ask a question. The adhesion effect is even more pronounced, and consumers are trained to move along in the process, to be able to select "next" and come one step closer to completion. Of course, as Jeff Jonas points out, the development of this "convenience" is driven as much by consumers' preference for convenience as by the sellers' motivation to complete more transactions (though of course, the two are related). We demand convenience, even though it comes with costs.
3. Bruce Schneier, "Fear and Convenience," in *Privacy in the Modern Age: The Search for Solutions*, ed. Marc Rotenberg, Julia Horwitz, and Jeramie Scott (New York: New Press, 2015), 202.
4. Glenn Greenwald, *Nothing to Hide: Edward Snowden, The NSA, and the US Surveillance State* (New York: Picador, 2015).
5. Dave Zirin and Andrew Tan Delli-Cicchi, "Fans Are the Target of Madison Square Garden's New Facial Recognition Technology," *Nation*, March 23, 2018, https:www.thenation.com/article/archive/fans-are-the-target-of-madison-square-gardens-new-facial-recognition-technology.

6. Manuel DeLanda, *A New Philosophy of Society* (London: Continuum, 2006).
7. Vance Packard, *The Hidden Persuaders* (New York: Longmans, Greene, 1957).
8. Packard, *Hidden Persuaders,* 7.
9. Packard, 96.
10. Packard, 15.
11. I am indebted to an anonymous reviewer for this insight.
12. Packard, 159.
13. Cass Sunstein, *Why Nudge? The Politics of Liberal Paternalism* (New Haven, CT: Yale University Press, 2015).
14. Sunstein, *Why Nudge?*, 14.
15. Sunstein, 7–8.
16. Sunstein, 7–8.
17. David Lyon, *The Culture of Surveillance* (Cambridge, MA: Polity, 2018), 39.
18. Aleecia McDonald and Lorrie Faith Craner, "The Cost of Reading Privacy Policies." *I/S: A Journal of Law and Policy for the Information Society* 4, no. 3 (2008): 540–565.
19. Joseph Turow, "America's Online Privacy: The System Is Broken," 2003, https://repository.upenn.edu/cgi/viewcontent.cgi?article=1411&context=asc_papers (accessed June 23, 2018).
20. Turow, "America's Online Privacy."
21. EU General Data Protection Regulation, 2018, https://ec.europa.eu/commission/priorities/justice-and-fundamental-rights/data-protection/2018-reform-eu-data-protection-rules_en.
22. Lee Rainie, "America's Complicated Feelings about Social Media in an Era of Privacy Concerns," http://www.pewresearch.org/fact-tank/2018/03/27/americans-complicated-feelings-about-social-media-in-an-era-of-privacy-concerns/.
23. Lee Rainie and Janna Anderson, "Trust Will Not Grow but Technology Usage Will Continue to Rise as a New Normal Sets In," 2017, http://www.pewinternet.org/2017/08/10/theme-3-trust-will-not-grow-but-technology-usage-will-continue-to-rise-as-a-new-normal-sets-in/.
24. Rainie and Anderson, "Trust Will Not Grow but Technology Usage Will Continue to Rise."
25. Schneier, "Fear and Convenience," 201n3.
26. Peter Adey, "Mobilities and Modulations: The Airport as a Difference Machine," in *Politics at the Airport*, ed. Mark B. Salter (Minneapolis: University of Minnesota Press, 2008), 145–157.
27. Adey, "Mobilities and Modulations," 157.
28. Mark Augé, *Non-Places: Introduction to an Anthropology of Supermodernity,* trans. John Howe (New York: Verso, 1995).
29. Gillian Fuller, "Welcome to Windows 2.0," in Salter, *Politics at the Airport,* 162–163.
30. Kevin Haggerty and Richard Ericson, "The Surveillant Assemblage," *British Journal of Sociology* 51, no. 4 (2000): 605–622.
31. Adey, "Mobilities and Modulations," 145n15.
32. D. Wood and S. Graham, "Permeable Boundaries in the Software Sorted Society: Surveillance and the Differentiation of Mobility," in *Mobile Technologies of the City,* ed. M. Sheller and J. Urry (London: Routledge, 2006).

33. Adey, "Mobilities and Modulations," 145.
34. Lyon, *Culture of Surveillance,* 45.
35. Michael Taussig, *Defacement: Public Secrecy and the Labor of the Negative* (Palo Alto, CA: Stanford University Press, 1999).
36. Lyon, *Culture of Surveillance,* 5nxxii (indicating the importance of imaginaries in studying surveillance).
37. Gary Marx, "Rocky Bottoms: Techno-fallacies of an Age of Information," *Political Sociology* 1, no. 1 (2007): 83–110.
38. Torin Monahan, *Surveillance in the Time of Insecurity* (New Brunswick, NJ: Rutgers University Press, 2010).
39. Frank Pasquale, "From Territorial to Functional Sovereignty: The Case of Amazon," *Law and Political Economy* (blog), https://lpeblog.org/ (accessed October 31, 2018).
40. Pasquale, "From Territorial to Functional Sovereignty."
41. Lyon, *Culture of Surveillance,* 32n23.
42. Nick Srnicek, *Platform Capitalism* (Malden, MA: Polity, 2017), 3.
43. Benjamin Muller, "Travelers, Borders, Dangers: Locating the Political at the Biometric Border," in Salter, *Politics at the Airport,* 136.
44. Presumably, affirmative answers to these questions would have generated follow-up questions.
45. Nikolas Rose, "Government and Control," *British Journal of Criminology* 40 (2000): 321–339, at 326.
46. In the years following publication of Rose's article, the increasing use of algorithms to process biofeedback and other health-related data has resulted in a loss of control over the processing/interpretation of the data on the part of physicians as well as patients. I am indebted to an anonymous reviewer for this observation.
47. Rose, "Government and Control," 326.
48. Colin Bennett, "Unsafe at Any Altitude: The Comparative Politics of No-Fly Lists," in Salter, *Politics at the Airport,* 55.
49. Lyon, *Culture of Surveillance,* 6.
50. It is worth noting that Amazon's cashier-less stores also watch shoppers via cameras hidden in the floor, so that the shoppers do not know they are being watched. See Abha Bhattarai and Drew Harwell, "Inside Amazon Go: The Camera-Filled Convenience Store That Watches You Back," *Washington Post,* January 22, 2018, https://www.washingtonpost.com/amphtml/news/business/wp/2018/01/22/inside-amazon-go-the-camera-filled-convenience-store-that-watches-you-back/.
51. Pramod Nayar, *Citizenship and Identity in the Age of Surveillance* (Delhi: Cambridge University Press, 2016), 150.
52. Nayar, *Citizenship and Identity,* 151.
53. Nayar, 151
54. Nayar, 139.
55. Nayar, 139.
56. Nayar, 151.
57. Nayar,150.
58. Nayar, 150.
59. Nayar, 150.

60. Lyon, *Culture of Surveillance*, 23.
61. Lyon, 23.
62. Schneier, "Fear and Convenience," 200.
63. Numerous commentators have pointed out that security protocols make us feel better regardless of whether we are *actually* safer—and that affective response provides a motivation for allowing inconvenience in this instance.
64. Cass Sunstein, "Terrorism and Probability Neglect," *Journal of Risk and Uncertainty* 26 (2003): 121–136.
65. For explanation of this term, see Monahan, *Surveillance in the Time of Insecurity*, 8n27.

Chapter 3. Technologies of Ubiquity and Obscurity

1. David Murakami Wood, "Vanishing Surveillance: Securityscapes and Ambient Government," March 20, 2014, http://pactac.net/2014/03/vanishing-surveillance-securityscapes-and-ambient-government/.
2. Jathan Sadowski and Frank Pasquale, "The Spectrum of Control: A Social Theory of the Smart City," *First Monday* 20, no. 7 (2015): 1–22, 9.
3. Sadowski and Pasquale, "Spectrum of Control," 10.
4. Nick Srnicek, *Platform Capitalism* 43(Cambridge, MA: Polity Press, 2017).
5. Srnicek,49.
6. Srnicek, 49.
7. Srnicek, 40.
8. Julie Cohen, "Law for the Platform Economy," *UC Davis Law Review* 51, no. 8 (2017): 133–205.
9. Online shoppers choose Target for shopping because they have already used the platform for another transaction; see Frank Pasquale, "From Territorial to Functional Sovereignty: The Case of Amazon," *Law and Political Economy* (blog), 2017, https://lpeblog.org/.
10. Pasquale, "From Territorial to Functional Sovereignty."
11. Cohen, "Law for the Platform Economy," 152.
12. Pasquale, "From Territorial to Functional Sovereignty."
13. Pasquale.
14. Aristotle, *Politics,* Book III.
15. Of course, there is no bright-line divide between physical and spatial. Online dating services arrange dates that take place in a physical location, and logistics platforms guide trucks driving on actual roads.
16. Matt Stevens, "Starbucks C.E.O. Apologizes After Arrests of 2 Black Men," *New York Times,* April 15, 2018, https://www.nytimes.com/2018/04/15/us/starbucks-philadelphia-black-men-arrest.html.
17. David Lyon, "Why Where You Are Matters: Mundane Mobilities, Transparent Technologies, and Digital Discrimination," in *Surveillance and Security,* ed. Torin Monahan (New York: Routledge, 2006), 222.
18. Of course, it is not absolutely irresistible. There are significant numbers of professionals who don't use Facebook. And this pattern varies by profession. Some academics, for example, rely on Facebook to share information about conferences and publications while some do not.

19. Sadowski and Pasquale, "Spectrum of Control."
20. Federal Trade Commission, *Data Brokers: A Call for Transparency and Accountability*, 2014, https://www.ftc.gov/system/files/documents/reports/data-brokers-call-transparency-accountability-report, iii.
21. Federal Trade Commission, *Data Brokers*, iv.
22. Torin Monahan, *Surveillance in the Time of Insecurity* (New Brunswick, NJ: Rutgers University Press, 2010), 51. In fact, nearly all of the computer hardware currently in use contains "baked-in" vulnerabilities in design, resulting from the construction of Intel chips themselves. Thus, the susceptibility to breach that Monahan describes is quite extensive. I am indebted to an anonymous reviewer for this insight.
23. Federal Trade Commission, *Data Brokers*, 68.
24. Federal Trade Commission, 61.
25. Chapter 22, California Business and Professions Code, https://leginfo.legislature.ca.gov/faces/codes_displayText.xhtml?division=8.&chapter=22.&lawCode=BPC (accessed April 1, 2019).
26. The Children's Online Privacy Protection Act of 1998 protects children under age thirteen from some privacy consequences of Internet use and allows parents to control what information can be collected about their children. 15 U.S.C. §§ 6501–6506.
27. Online Privacy Bill of Rights, https://obamawhitehouse.archives.gov/sites/default/files/privacy- final.pdf (accessed April 1, 2019).
28. Paul Bischoff, "What Is the Consumer Privacy Bill of Rights?," https://www.comparitech.com/blog/vpn-privacy/consumer-privacy-bill-of-rights/ (accessed April 1, 2019).
29. Martin Kuhn, *Federal Dataveillance: Implications for Constitutional Privacy Protections* (New York: LFB Scholarly Publishing, 2007).
30. Pramod Nayar, *Citizenship and Identity in the Age of Surveillance* (Delhi: Cambridge University Press, 2016), 87.
31. Federal Trade Commission, *Data Brokers*, 14.
32. Federal Trade Commission, 14.
33. Frank Pasquale, "The Algorithmic Self," *Hedgehog Review* 17, no. 1 (Spring 2015): 4.
34. Louis Althusser, *Lenin and Philosophy and Other Essays* (New York: Verso, 1971), 163.
35. Federal Trade Commission, *Data Brokers*, 53.
36. Gillian White, "A Cybersecurity Breach at Equifax Left Pretty Much Everyone's Financial Data Vulnerable," *Atlantic*, September 7, 2017, https://www.theatlantic.com/business/archive/2017/09/equifax-cybersecurity-breach/539178/.
37. Daniel Solove and Danielle Citron, "Risk and Anxiety: A Theory of Data Breach Harms," *Texas Law Review* 96 (2018): 737–786, at 745.
38. Solove and Citron, "Risk and Anxiety," 745.
39. Solove and Citron, 745.
40. Solove and Citron, 750.
41. Solove and Citron, 750.
42. Solove and Citron, 750.
43. How many is another question.
44. Cohen, "Law for the Platform Economy," 24.

45. Cohen, 16.
46. Cohen, 25.
47. Cohen, 13.
48. Cohen, 13.
49. Sadowski and Pasquale, "Spectrum of Control," 14.
50. Robinson Meyer, "Everything We Know about Facebook's Secret Mood Manipulation Experiment," *Atlantic,* June 28, 2014, https://www.theatlantic.com/technology/archive/2014/06/everything-we-know-about-facebooks-secret-mood-manipulation-experiment/373648/.
51. Gary Marx, "Rocky Bottoms: Techno-fallacies of an Age of Information," *Political Sociology* 1, no. 1 (2007): 83–110, at 40.

Chapter 4. The Internet of Things and the Smart City

1. David Lyon, *The Culture of Surveillance* (Boston: Polity, 2018), 50.
2. Jathan Sadowski and Frank Pasquale, "The Spectrum of Control: A Social Theory of the Smart City," *First Monday* 20, no. 7 (July 6, 2015), 1.
3. Sadowski and Pasquale, "Spectrum of Control," 1.
4. Adam Greenfield, *Against the Smart City* (New York: Do Projects, 2013), 119.
5. Rob Kitchin, "The Real-Time City? Big Data and Smart Urbanism," *GeoJournal* 79 (2014): 1–14, at 8.
6. Christine Rosen, "The Machine and the Ghost," *New Republic,* https://newrepublic.com/article/104874/rosen-verbeek-technology-morality-intelligence (accessed April 6, 2017).
7. Rosen, "The Machine and the Ghost."
8. Sadowski and Pasquale, "Spectrum of Control," 2.
9. Sadowski and Pasquale, 2.
10. Adam Greenfield, "Adam Greenfield on Connected Things & Civic Responsibilities in the Networked City," June 21, 2011, https://www.youtube.com/watch?v=7C9kgLDxCS4.
11. And other cities as well. See Sadowski and Pasquale, "Spectrum of Control," 4.
12. Sadowski and Pasquale, 12.
13. Sadowski and Pasquale, 3.
14. Sadowski and Pasquale, quoting Wacquant, 3.
15. Sadowski and Pasquale, 5.
16. Greenfield, *Against the Smart City,* 741.
17. Gary Marx, "Rocky Bottoms: Techno-fallacies of an Age of Information," *Journal of International Political Sociology* 1, no. 1 (2007): 83–110.
18. Sadowski and Pasquale, "Spectrum of Control," 5.
19. Sadowski and Pasquale, quoting Bogard, 7.
20. Sadowski and Pasquale, quoting Bogard, 7.
21. Pramod Nayar, *Citizenship and Identity in the Age of Surveillance* (Delhi: Cambridge University Press, 2015), 95.
22. Sadowski and Pasquale, "Spectrum of Control," 7.
23. Greenfield, *Against the Smart City,* 865.
24. Greenfield, 759.
25. Greenfield, 780.

26. Greenfield, 780.
27. McKinsey Global Institute, *Smart Cities: Digital Solutions for a More Livable Future*, 2018, https://www.mckinsey.com/~/media/mckinsey/industries/capital%20projects%20and%20infrastructure/our%20insights/smart%20cities%20digital%20solutions%20for%20a%20more%20livable%20future/mgi-smart-cities-full-report.ashx.
28. McKinsey Global Institute, *Smart Cities.*
29. I am indebted to an anonymous reviewer for this insight.
30. Greenfield makes this point by examining "high modernist" architecture in the same time period and highlighting some of the functionalist assumptions at work there; Greenfield, *Against the Smart City,* chapter 13.
31. Gray Brechin, *Imperial San Francisco: Urban Power, Earthly Ruin* (Berkeley: University of California Press, 1998), xxiv.
32. Brechin, *Imperial San Francisco,* 110.
33. Victor Valle, *City of Industry: Genealogies of Power in Southern California* (New Brunswick, NJ: Rutgers University Press, 2009).
34. Robert Caro, *The Power Broker* (New York: Vintage, 1974).
35. The most infamous example, cited by Winner and others, was Moses's decision to build low-clearance bridges over borough expressways in order to prevent city buses carrying persons of color from accessing Jones Beach and other public facilities. Caro also notes that parks built alongside high-speed roadways functioned to segregate people in the same way: the residents of one neighborhood could easily reach the park, while those on the other side had to cross a dangerous road to get there. Incredibly, Moses is also said to have maintained low water temperatures in public pools out of the belief that African American residents inherently disliked cold water (!). Caro, *The Power Broker*, 318.
36. Lyon, *Culture of Surveillance,* 90.
37. Langdon Winner, *The Whale and the Reactor: A Search for Limits in the Age of High Technology* (Chicago: University of Chicago Press, 1986).
38. Sadowski and Pasquale, "Spectrum of Control," 14.
39. Greenfield, *Against the Smart City*, 65.
40. Greenfield, 65.
41. Greenfield,393.
42. Greenfield,283.
43. Sadowski and Pasquale, "Spectrum of Control," 7.
44. Sadowski and Pasquale, 7.
45. Nayar, *Citizenship and Identity in the Age of Surveillance*, 6.
46. Nayar, 6.
47. Marx, "Rocky Bottoms," 270.
48. See Lyon, *Culture of Surveillance*, 84.
49. Lyon, 96.
50. One exception might be a high-tech, luxury development such as Hudson Yards in New York City, where residents have been attracted by the overall grandeur of a community with numerous smart-tech features that is not, strictly speaking, a smart city *ab initio.* Ironically, Hudson Yards creator Stephen Ross has been compared to Robert Moses in terms of his development vision, though not for any of Moses's malevolent aims. Carl Swanson, "The Only

Man Who Could Build Oz," 2019, http://nymag.com/intelligencer/2019/02/stephen-ross-hudson-yards.html.

51. Greenfield, *Against the Smart City*, 443.
52. Greenfield, 444.
53. Hannah Arendt, *The Human Condition* (Chicago: University of Chicago Press, 1998).
54. Lyon, *Culture of Surveillance*, 31, 41.
55. Marx, "Rocky Bottoms," 270.
56. Greenfield, *Against the Smart City*, 808.
57. Torin Monahan, "Editorial: Algorithmic Fetishism," *Surveillance and Society* 16, no. 1 (2018): 1–5, at 2.
58. Torin Monahan, "The Image of the Smart City: Surveillance Protocols and Social Inequality," in *Handbook of Cultural Security*, ed. Y. Watanabe (Cheltenham, UK: Edward Elgar, 2018), 210.
59. Monahan, "Editorial: Algorithmic Fetishism," 2.
60. Kitchin, "The Real-Time City?," 9.
61. Greenfield highlights his critique of technocratic urban planning by describing the RAND studies of the 1970s that attempted to minimize response times for firefighters in New York. The study was intended to determine the ideal placement of fire stations throughout the city for rapid response. In the end, the RAND project devolved into political infighting and the recommendations actually made the existing system worse for some inner-city New York neighborhoods. Greenfield, *Against the Smart City*, 528–529.
62. Greenfield, 741.
63. Kitchin, "The Real-Time City?," 10.
64. Kitchin, 12.
65. Greenfield, *Against the Smart City*, 1113.
66. Greenfield, 283.
67. Greenfield, 980. The importance of a public sphere for a functioning democracy has been emphasized by a number of theorists in recent years. See, especially, Jürgen Habermas, *The Structural Transformation of the Public Sphere: An Inquiry into a Category of Bourgeois Society* (Cambridge, MA: MIT Press, 1991). Habermas notes the eclipse of the bourgeois public sphere in the late nineteenth and early twentieth centuries by industrial capitalism, mass advertising, and the rise of the consumer society. Nancy Fraser offers a more robust and complex description of the public sphere as a persisting feature of late modernity, featuring publics and counter-publics that vary in strength. In a sense, the smart city as planned and administered in the twenty-first century has imperiled any notion of a viable public sphere by monopolizing both physical and discursive space. Nancy Fraser, "Rethinking the Public Sphere: A Contribution to the Critique of Actually Existing Democracy," *Social Text* 25/26 (1990): 56–80.
68. Greenfield, *Against the Smart City*, 753.
69. "Megacities: Urban Future, the Emerging Complexity," video, https://theintercept.com/2016/10/13/pentagon-video-warns-of-unavoidable-dystopian-future-for-worlds-biggest-cities/ (accessed May 6, 2019). I am indebted to Gray Brechin for directing me to this film.
70. For a discussion of the trade-offs involved in joining a gated community,

see Monahan, *Surveillance in the Time of Insecurity*, chapter 6 ("Residential Fortifications).

Chapter 5. Avenues of Resistance

1. This term originates with Torin Monahan, *Surveillance in the Time of Insecurity* (New Brunswick, NJ: Rutgers University Press, 2010).
2. Gary Marx, "A Tack in the Shoe: Neutralizing and Resisting the New Surveillance," *Journal of Social Issues* 59, no. 2 (2003): 369–390, at 384.
3. The state trials resulted in acquittal of the officers involved; a subsequent federal trial for criminal civil rights violations led to two of the officers' convictions. Douglas Linder, "The Rodney King Beating Trials," 2001, https://web.archive.org/web/20091203041211/http://jurist.law.pitt.edu/trials24.htm.
4. NYPD officer Daniel Pantaleo was fired five years after a cell phone video showed him placing unarmed, nonviolent African American civilian Eric Garner in a chokehold. Garner died as a result of the incident, but Pantaleo was not indicted, so he faced no criminal prosecution; Ashley Southall, "Daniel Pantaleo, Officer Who Held Eric Garner in Chokehold, Is Fired," *New York Times,* August 19, 2019, https://www.nytimes.com/2019/08/19/nyregion/daniel-pantaleo-fired.html. In 2015 North Charleston, South Carolina, police officer Michael Slager shot and killed unarmed African American civilian Walter Scott. A video recording shows that Slager shot Scott in the back as Scott was running away. Officer Slager was prosecuted in state and federal court and ultimately entered a federal plea agreement that resulted in a twenty-year prison sentence; Michael Boroff, "Slager Gets 19 to 24 Years in Fatal Shooting of Walter Scott," *New York Daily News,* December 7, 2017, https://www.nydailynews.com/news/crime/michael-slager-forgiven-mom-walter-scott-20-years-article-1.3683402?barcprox=true.
5. Marx, "A Tack in the Shoe," 384; see also Richard V. Ericson and Kevin Haggerty, *Policing the Risk Society* (Oxford: Clarendon, 1997), 56–58.
6. Richard W. Evans, "'The Footage Is Decisive': Applying the Thinking of Marshall McLuhan to CCTV and Police Misconduct." *Surveillance and Society* 13, no. 2 (2015): 215–232.
7. Drew Harwell, "Doorbell-Camera Firm Ring Has Partnered with 400 Police Forces, Extending Surveillance Concerns," *Washington Post,* August 28, 2019, https://www.washingtonpost.com/technology/2019/08/28/doorbell-camera-firm-ring-has-partnered-with-police-forces-extending-surveillance-reach/.
8. John Herrman, "All the Crime, All the Time: How *Citizen* Works," *New York Times,* March 17, 2019, https://www.nytimes.com/2019/03/17/style/citizen-neighborhood-crime-app.html.
9. Glenn Greenwald, *No Place to Hide: Edward Snowden, the NSA, and the US National Security State* (New York: Metropolitan, 2014), 6. See also Chris Jaikaran, "Encryption: Frequently Asked Questions," 2016, https://fas.org/sgp/crs/misc/R44642.pdf.
10. Greenwald, *No Place to Hide,* 8.
11. Some would disagree with the characterization of Snowden as a whistleblower. The federal government claimed Snowden had committed a crime

by disclosing classified information, whereas a whistleblower is generally thought to expose illegal activity *by the government*. I leave that judgment to the reader; my point is to highlight the significance of the information disclosed. Whatever one thinks of the legality of the leak, it provided the public with information about a staggering amount of data collected from the general public. The data collection was made possible by the cooperation of private service providers and social media platform owners.

12. Greenwald, *No Place to Hide*, 10.
13. Ellen Nakashima and Barton Gellman, "As Encryption Spreads, US Worries About Access to Data for Investigations," *Washington Post*, August 10, 2015, https://www.washingtonpost.com/world/national-security/as-encryption-spreads-us-worries-about-access-to- data-for-investigations/2015/04/10/7c1c7518-d401-11e4-a62f-ee745911a4ff_story.html.
14. Josephine Wolff, "The Most Shocking Thing About Encrypted Email Being Vulnerable Is That Anyone Still Uses Encrypted Email," *Salon*, May 15, 2018, https://slate.com/technology/2018/05/pgp-and-s-mime-are-vulnerable-but-also-no-one-used-them-anyway.html.
15. Kevin Werbach and Nicholas Cornell, "Contracts Ex Machina," *Duke Law Journal* 67 (2017): 314n3.
16. Werbach and Cornell, "Contracts Ex Machina," 324.
17. The rules specify that the blockchain will be halved periodically so that it does not grow too large.
18. Kevin Werbach, "Trust but Verify: Why Blockchain Needs the Law," *Berkeley Technology Law Journal* 33 (2018): 487–550, at 500.
19. Werbach and Cornell, "Contracts Ex Machina," 323.
20. Werbach and Cornell, 355; *Lochner v. New York*, 198 U.S. 45 (1905). Werbach suggests that most people who use smart contracts do so for practical rather than ideological reasons: it helps to solve problems and make transactions easier; Werbach, "Trust but Verify," 508.
21. Werbach and Cornell, "Contracts Ex Machina," 15.
22. Werbach, "Trust but Verify," 507.
23. Most notoriously, the online illegal drug marketplace Silk Road saw $200 million in sales before it was shut down and its operator was sentenced to life in prison; Sam Thielman, "Silk Road Operator Ross Ulbricht Sentenced to Life in Prison," *Guardian*, May 29, 2015, https://www.theguardian.com/technology/2015/may/29/silk-road-ross-ulbricht-sentenced.
24. Werbach and Cornell, "Contracts Ex Machina," 376.
25. See, for example, Section 215 of the Uniting and Strengthening America by Providing Appropriate Tools Required to Intercept and Obstruct Terrorism Act of 2001 ("USA Patriot Act"), 50 U.S.C. § 1861 (2012). This provision is known as the "business records" provision and must be reauthorized periodically.
26. Mark Sundeen, *The Man Who Quit Money* (New York: Riverhead, 2011).
27. James M. Harding, *Performance, Transparency, and the Culture of Surveillance* (Ann Arbor: University of Michigan Press, 2018), 231.
28. Shaun Walker, "Petr Pavlensky in Court after Setting Fire to Lubyanka," *Guardian*, April 28, 2016, https://www.theguardian.com/world/2016/apr/28/petr-pavlensky-appears-court-russia-setting-fire-to-lubyanka-protest.

29. The website for the exhibit is https://theshed.org/program/63-manual-override (accessed January 23, 2020).
30. The installation is entitled "Shadow Stalker," and it was created by Lynn Hershman Leeson. A description can be found at https://www.lynnhershman.com/project/shadow-stalker/ (accessed January 23, 2020). Regarding identification errors, precisely that kind of mistake led to the rendition and torture of Murat Kurnaz, a Turkish citizen living in Germany who was erroneously labeled a terrorist because of his association with an alleged suicide bomber. When Kurnaz eventually learned of the reason for his imprisonment, he was able to show that the "suicide bomber"—a former coworker of his—was alive and well and had no ties to terrorism. Murat Kurnaz, *Five Years of My Life: An Innocent Man in Guantanamo* (New York: Palgrave MacMillan, 2008).
31. Louis Althusser, *Lenin and Philosophy and Other Essays* (New York: Verso, 1971), 163.

Chapter 6. Conclusion: Technology, Surveillance, and the Social Contract

1. Simon Chesterman, *One Nation under Surveillance: A New Social Contract to Defend Freedom without Sacrificing Liberty* (New York: Oxford University Press, 2011), 249.
2. John Locke, *Two Treatises of Government,* 1689, https://oll.libertyfund.org/pages/john-locke-two-treatises-1689; Thomas Hobbes, *Leviathan,* 1651, https://www.gutenberg.org/files/3207/3207-h/3207-h.htm.
3. For the use of this term by the political philosopher John Rawls, see Leif Wenar, "John Rawls," *The Stanford Encyclopedia of Philosophy* (Spring 2017), ed. Edward N. Zalta, https://plato.stanford.edu/archives/spr2017/entries/rawls/.
4. Chesterman, *One Nation under Surveillance,* 249.
5. This idea was developed by Antonio Gramsci in his study of the process by which a social group becomes "hegemonic." Gramsci claimed that a more stable ruling order could be created when a social group constructed a worldview justifying its own rule that was in turn accepted by the population to be ruled. This, for Gramsci, was hegemony, and the concept has subsequently been applied in international relations theory, political activism, and cultural studies, among other contexts. Antonio Gramsci, *Selections from the Prison Notebooks of Antonio Gramsci* (New York: International Publishers, 1971).
6. Torin Monahan, *Surveillance in the Time of Insecurity* (New Brunswick, NJ: Rutgers University Press, 2010), 23.
7. Chesterman, *One Nation under Surveillance,* 255.
8. David Lyon, *The Culture of Surveillance: Watching as a Way of Life* (Cambridge, MA: Polity, 2018), 173.
9. Throughout this book I have followed the lead of other scholars studying contemporary power relations who emphasize that corporate actors as well as state actors engage in governance. Both seek to alter the behavior of individuals and groups in order to facilitate social control, albeit for different ends.

10. The Contracts Clause of the Constitution, Article I, section 10, was held by the Supreme Court to protect the expectations of contracting parties against government infringement, even when that infringement took the form of legislative acts. *Fletcher v. Peck,* 10 U.S. 87 (1810), and *Sturges v. Crowninshield,* 17 U.S. 122 (1819), were two early cases so to hold. Eventually the doctrine gave way to a narrower conception of private rights that had to be balanced against the public good (see *Charles River Bridge v. Warren Bridge,* 36 U.S. 420 (1837).
11. Robert Caro, *The Power Broker* (New York: Alfred A. Knopf, 1974).
12. Mark Andrejevic, "'Securitainment'" in the Post-9/11 Era," *Continuum: Journal of Media and Cultural Studies* 25, no. 2 (2011): 165–175, at 168.
13. Michel Foucault, *Discipline and Punish: The Birth of the Prison* (New York: Pantheon, 1977).
14. Andrejevic, "'Securitainment,'" 168.
15. Lyon, *Culture of Surveillance*; Monahan, *Surveillance in the Time of Insecurity.*
16. Andrejevic, "'Securitainment,'" 169.
17. Andrejevic, 167.
18. David Campbell makes this connection between external enemies and internal US consumer choices by referencing the popularity of sport utility vehicles (SUVs). Because SUVs require large amounts of gasoline, the public demand for them and the resulting demand for foreign oil generates foreign policy choices that construct residents of oil-rich nations as enemies of America and of freedom. Consumer behavior produces an exoticized enemy Other by envisioning "borderlands" that are "conventionally understood as distant, wild places of insecurity where foreign intervention will be necessary to ensure domestic interests are secured." David Campbell, "The Biopolitics of Security: Oil, Empire, and the Sport Utility Vehicle," *American Quarterly* 57, no. 3 (2005): 943–972, at 946.
19. This term comes from Isaiah Berlin's essay "Two Concepts of Liberty." Isaiah Berlin, *Four Essays on Liberty* (Oxford: Oxford University Press, 1969), 118–172.
20. Berlin, *Four Essays on Liberty,* 125.
21. *Korematsu v. US,* 323 U.S. 214 (1944).
22. See for example, Justice Thomas's dissent in *Hamdi v. Rumsfeld,* 542 U.S. 507 (2004). Though the Court ruled in favor of protecting civil liberties in a few instances, it has applied a hands-off approach to detainee suits and related matters over the past ten years.
23. For an explanation of this doctrine, see William Weaver and Robert Pallitto, "State Secrets and Executive Power," *Political Science Quarterly* 120, no. 10 (2005): 85–112.
24. In this regard, see Gary Marx, "A Tack in the Shoe: Neutralizing and Resisting the New Surveillance," *Journal of Social Issues* 59, no. 2 (2003): 369–390.
25. Lyon, *Culture of Surveillance*; Frank Pasquale, "The Algorithmic Self," *Hedgehog Review* 17, no. 1 (Spring 2015).
26. Pasquale, "The Algorithmic Self."
27. Lyon, *Culture of Surveillance,* 23.
28. Lyon,185.
29. Chantal Mouffe and Ernesto Laclau, *Hegemony and Socialist Strategy* (London: Verso, 1985).

30. Hannah Arendt, *The Human Condition* (Chicago: University of Chicago Press, 1958),169.
31. Michael Rosen, *On Voluntary Servitude: False Consciousness and the Theory of Ideology* (Cambridge, MA: Harvard University Press, 1996), 1.
32. Lyon, *Culture of Surveillance,* 121.
33. "Text, President Bush Addresses the Nation," *Washington Post,* September 20, 2001, http://www.washingtonpost.com/wp-srv/nation/specials/attacked/transcripts/bushaddress_092001.html. Bush said, "I ask your continued participation and confidence in the American economy." On another occasion, the president referenced airline travel specifically, saying, "And one of the great goals of this nation's war is to restore public confidence in the airline industry, is to tell the traveling public, 'Get on board. Do your business around the country. Fly and enjoy America's great destination spots. Go down to Disney World in Florida, take your families and enjoy life the way we want it to be enjoyed.'" "Text, Bush on Airline Safety Measures," *Washington Post,* September 27, 2001, http://www.washingtonpost.com/wp-srv/nation/specials/attacked/transcripts/bush_092701.html.
34. Monahan is quoting a Homeland Security official who suggested that department stores—and Wal-Mart in particular—should be a high priority for protection in a security crisis. Monahan documents this neoliberal tendency to enlist state power and resources to protect business interests. Monahan, *Surveillance in the Time of Insecurity,* 40.
35. Jürgen Habermas, *The Structural Transformation of the Public Sphere* Cambridge, MA: MIT Press, 1989), 192.
36. Michael Sandel, "Reply to Critics," in *Debating Democracy's Discontent: Essays on American Politics, Law, and Public Philosophy.,* ed. Anita Allen and Milton Regan Jr. (New York: Oxford University Press, 1998), 320.
37. Jean B. Elshtain and Christopher Beem, "Can This Republic Be Saved?," in Allen and Regan, *Debating Democracy's Discontent,* 197.
38. Elshtain and Beem, "Can This Republic Be Saved?," 197.
39. Elshtain and Beem, 198.
40. Robert Verchick, *Facing Catastrophe: Environmental Action for a Post-Katrina World* (Cambridge, MA: Harvard University Press, 2010), 208. Similarly, Mike Davis notes how consumer choices to live on fault lines or in brushfire corridors have exacerbated the effects of disasters such as earthquakes and wildfires. The effects are distributed across the population at large, and then they are made to seem inevitable when in fact they resulted from human action. Mike Davis, *The Ecology of Fear: Los Angeles and the Imagination of Disaster* (New York: Vintage, 1999).
41. Verchick, *Facing Catastrophe,* 195.
42. Verchick,236–239.
43. Verchick, 236–239.
44. Verchick, 236–239; Joseph Singer, "Something Important in Humanity," *Harvard Civil Rights-Civil Liberties Law Review* 37 (2002): 103–130.
45. See Morton Horwitz, *The Transformation of American Law 1780–1860* (Cambridge, MA: Harvard University Press, 1977).
46. A technical term in contract law referring to what each a party gives up and receives in an exchange of promises forming a contract.

47. See *Williams v. Walker-Thomas Furniture Co.*, 350 F.2d 445 (D.C. Cir. 1965), the landmark case on unconscionable contracts.
48. Singer uses the more poignant example of a society whose flourishing was made possible by an endlessly tortured child. Most of us (hopefully) would make a values-based objection to that arrangement, whatever our preferences.
49. Don Mitchell, "The S.U.V. Model of Citizenship: Floating Bubbles, Buffer Zones, and the Rise of the 'Purely Atomic' Individual," *Political Geography* 24 (2005): 77–100.
50. Campbell, "Biopolitics of Security," 943–972.
51. Campbell, "Biopolitics of Security," 966.
52. Campbell, 967.
53. Anita Allen, "An Ethical Duty to Protect One's Own Information Privacy?," *Alabama Law Review* 64, no. 4 (2013): 845–866.
54. Allen, "An Ethical Duty," 848.
55. Allen, 845.
56. Allen, 855.
57. Allen, 850.
58. Allen, 864.
59. Allen, 860.
60. Allen, 865.
61. Federal Trade Commission, *Data Brokers: A Call for Transparency and Accountability*, 2014, https://www.ftc.gov/system/files/documents/reports/data-brokers-call-transparency-accountability-report-federal-trade-commission-may-2014/140527databrokerreport.pdf.
62. Slavoj Žižek, *The Sublime Object of Ideology* (New York: Verso, 1989), 33.

BIBLIOGRAPHY

Adey, Peter. "Secured and Sorted Mobilities: Examples from the Airport." *Surveillance and Society* 1, no. 4 (2004): 500–519.

Adorno, Theodor. *Minima Moralia: Reflections from Damaged Life.* New York: Verso, 2006.

Allen, Anita. "An Ethical Duty to Protect One's Own Information Privacy?" *Alabama Law Review* 64, no. 4 (2013): 845–866.

———. *Unpopular Privacy: What Must We Hide?* New York: Oxford University Press, 2011.

Allen, Anita, and Milton Regan Jr., eds. *Debating Democracy's Discontent : Essays on American Politics, Law, and Public Philosophy.* New York: Oxford University Press, 1998.

Althusser, Louis. *Lenin and Philosophy and Other Essays.* New York: Verso, 1971.

Andreassen, Cecille S., J. Billieux, M. D. Griffiths, D. J. Kuss, Z. Demetrovics, E. Mazzoli, and S. Pallesen. "The Relationship between Addictive Use of Social Media and Video Games and Symptoms of Psychiatric Disorders: A Large-Scale Cross-Sectional Study." *Psychology of Addictive Behaviors* 30, no.2 (2016): 252–262.

Andrejevic, Mark. "'Securitainment'" in the Post-9/11 Era." *Continuum: Journal of Media and Cultural Studies* 25, no. 2 (2011): 165–175.

Arendt, Hannah. *The Human Condition.* Chicago: University of Chicago Press, 1958.

Augé, Mark. *Non-Places: Introduction to an Anthropology of Supermodernity.* Translated by John Howe. New York: Verso, 1995.

Benjamin, Walter. *Illuminations.* New York: Harcourt, Brace & Jovanovich, 1968.

Bennett, Colin. "Unsafe at Any Altitude: The Comparative Politics of No-Fly Lists." In *Politics at the Airport,* edited by Mark B. Salter. Minneapolis: University of Minnesota Press, 2008.

Berlin, Isaiah. "Two Concepts of Liberty." In *Four Essays on Liberty,* 118–172. Oxford: Oxford University Press, 1969.

Boorstin, Daniel. *The Republic of Technology: Reflections on Our Future Community.* New York: Harper & Row, 1978.

Brechin, Gray. *Imperial San Francisco: Urban Power, Earthly Ruin.* Berkeley: University of California Press, 1999.

Campbell, David. "The Biopolitics of Security: Oil, Empire, and the Sport Utility Vehicle." *American Quarterly* 57, no. 3 (2005): 943–972.

Campbell, Nancy. "Everyday Insecurities: The Microbehavioral Politics of Intrusive Surveillance." In *Surveillance and Security,* edited by Torin Monahan. New York: Routledge, 2006.

Caro, Robert. *The Power Broker.* New York: Vintage, 1974.

Chesterman, Simon. *One Nation under Surveillance: A New Social Contract to Defend Freedom without Sacrificing Liberty.* New York: Oxford University Press, 2011.

Cohen, Julie. "Law for the Platform Economy." *UC Davis Law Review* 51, no. 8 (2017): 133–205.

Davis, Mike. *The Ecology of Fear: Los Angeles and the Imagination of Disaster.* New York: Vintage, 1999.
DeLanda, Manuel. *A New Philosophy of Society.* London: Continuum, 2006.
———. *War in the Age of Intelligent Machines.* Brooklyn, NY: Zone, 1991.
Deleuze, Gilles, and F. Guattari. *Anti-Oedipus.* London: Continuum, 2004.
Doctorow, Cory. "Privacy, Public Health, and the Moral Hazard of Surveillance." *Guardian,* May 21, 2013.
Duhigg, Charles. "How Companies Learn Your Secrets (Psst, You in Aisle 5)." *New York Times Magazine,* February 16, 2012, MM30.
Elshtain, Jean Bethke, and Christopher Beem. "Can This Republic Be Saved?" In *Debating Democracy's Discontent*, edited by Anita L. Allen and Milton C. Regan. New York: Oxford University Press, 1998.
Ericson, Richard V., and Kevin D. Haggerty. *Policing the Risk Society.* Oxford: Clarendon, 1997.
Eubanks, Virginia. "Technologies of Citizenship: Surveillance and Political Learning in the Welfare System." In *Surveillance and Security,* edited by Torin Monahan. New York: Routledge, 2006.
Evans, Richard W. "'The Footage Is Decisive': Applying the Thinking of Marshall McLuhan to CCTV and Police Misconduct." *Surveillance and Society* 13, no. 2 (2015): 215–232.
Federal Trade Commission. *Data Brokers: A Call for Transparency and Accountability.* 2014. https://www.ftc.gov/system/files/documents/reports/data-brokers-call-transparency-accountability-report-federal-trade-commission-may-2014/140527databrokerreport.pdf.
Foucault, Michel. "The Body of the Condemned." In *The Foucault Reader*, edited by Paul Rabinow. New York: Pantheon, 1974.
Fraser, Nancy. "Rethinking the Public Sphere: A Contribution to the Critique of Actually Existing Democracy." *Social Text* 25/26 (1990): 56–80.
Fuller, Gillian. "Welcome to Windows 2.0." In *Politics at the Airport*, edited by Mark B. Salter. Minneapolis: University of Minnesota Press, 2008.
Gandy, Oscar. *Coming to Terms with Chance: Engaging Rational Discrimination and Cumulative Disadvantage.* Burlington, VT: Ashgate, 2009.
———. *The Panoptic Sort: A Political Economy of Personal Information.* Boulder, CO: Westview, 1993.
Gramsci, Antonio. *Selections from the Prison Notebooks of Antonio Gramsci.* New York: International Publishers, 1971.
Greenfield, Adam. *Against the Smart City.* New York: Do Projects, 2013.
———. *Radical Technologies: The Design of Everyday Life.* New York: Verso, 2017.
Greenwald, Glenn. *No Place to Hide: Edward Snowden, the NSA, and the US Surveillance State.* New York: Picador, 2015.
Griffiths, Mark. "A 'Components' Model of Addiction within a Biopsychosocial Framework." *Journal of Substance Abuse* 10, no. 4 (2008): 191–197.
Habermas, Jurgen. "Discourse Ethics: Notes on a Program of Philosophical Justification." In *Moral Consciousness and Communicative Action.* Translated by Christian Lenhart and Shierry Weber Nicholson. Cambridge, MA: MIT Press, 1990.
———. *The Structural Transformation of the Public Sphere: An Inquiry into a Category of Bourgeois Society.* Cambridge, MA: MIT Press, 1989.

Haggerty, Kevin, and Richard Ericson. "The Surveillant Assemblage." *British Journal of Sociology* 51, no. 4 (2000): 605–622.

Haraway, Donna. *Simians, Cyborgs, and Women: The Reinvention of Nature.* New York: Routledge, 1991.

Harding, James. *Performance, Transparency, and the Cultures of Surveillance.* Ann Arbor: University of Michigan Press, 2018.

Harwell, Drew. "Doorbell-Camera Firm Ring Has Partnered with 400 Police Forces, Extending Surveillance Concerns." *Washington Post,* August 28, 2019. https://www.washingtonpost.com/technology/2019/08/28/doorbell-camera-firm-ring-has-partnered-with-police-forces-extending-surveillance-reach/.

Heidegger, Martin. *The Question Concerning Technology, and Other Essays.* New York: Harper Torchbooks, 1977.

Herrman, John. "All the Crime, All the Time: How *Citizen* Works." *New York Times,* March 17, 2019. https://www.nytimes.com/2019/03/17/style/citizen-neighborhood-crime-app.html.

Hobbes, Thomas. *Leviathan.* New York: Penguin, 2017.

Holland, Eugene. *Deleuze and Guattari's Anti-Oedipus: An Introduction to Schizoanalysis.* New York: Routledge, 1999.

Horwitz, Morton. *The Transformation of American Law 1870–1960: The Crisis of Legal Orthodoxy.* New York: Oxford University Press, 1977.

Igo, Sarah. *The Known Citizen: A History of Privacy in Modern America.* Cambridge, MA: Harvard University Press, 2018.

Ihde, Don. *Philosophy of Technology: An Introduction.* New York: Paragon House, 2014.

Jaikaran, Chris. "Encryption: Frequently Asked Questions." 2016. https://fas.org/sgp/crs/misc/R44642.pdf.

Kitchin, Rob. "The Real-Time City? Big Data and Smart Urbanism." *GeoJournal* 79 (2014): 1–14.

Kurnaz, Murat. *Five Years of My Life: An Innocent Man in Guantanamo.* New York: Palgrave MacMillan, 2008.

Laclau, Ernesto, and Chantal Mouffe. *Hegemony and Socialist Strategy: Toward a Radical Democratic Politics.* New York: Verso, 1985.

Locke, John. *Two Treatises of Government.* New York: Cambridge University Press, 2013.

Lyon, David. *The Culture of Surveillance.* Cambridge, MA: Polity, 2013.

Madden, Mary. "Public Perceptions of Privacy and Security in the Post-Snowden Era." 2014. http://www.pewinternet.org/2014/11/12/public-privacy-perceptions/.

Marcuse, Herbert. *One-Dimensional Man: Studies in the Ideology of Advanced Industrial Society.* New York: Routledge, 2002.

Marr, Bernard. "A Complete Beginner's Guide to Blockchain." *Forbes,* January 24, 2017. https://www.forbes.com/sites/bernardmarr/2017/01/24/a-complete-beginners-guide-to-blockchain/#1fa61c896e60.

Marx, Gary. "Rocky Bottoms: Techno-fallacies of an Age of Information." *Political Sociology* 1, no. 1 (2007): 83–110.

———. "Surveillance Studies." *International Encyclopedia of Social and Behavioral Sciences,* 2nd ed. Edited by James D. Wright. Amsterdam: Elsevier, 2015.

McDonald, Aleecia, and Lorrie Faith Craner. "The Cost of Reading Privacy Policies."

I/S: A Journal of Law and Policy for the Information Society 4, no. 3 (2008): 540–565.

McKinsey Global Institute. *Smart Cities: Digital Solutions for a More Livable Future.* 2018.https://www.mckinsey.com/~/media/mckinsey/industries/capital%20projects%20and%20infrastructure/our%20insights/smart%20cities%20digital%20solutions%20for%20a%20more%20livable%20future/mgi-smart-cities-full-report.ashx.

Mitchell, Don. "The S.U.V. Model of Citizenship: Floating Bubbles, Buffer Zones, and the Rise of the 'Purely Atomic' Individual." *Political Geography* 24 (2005): 77–100.

Monahan, Torin. "Algorithmic Fetishism." *Surveillance and Society* 16, no. 1 (2018): 1–5.

———. "The Image of the Smart City: Surveillance Protocols and Social Inequality." In *Handbook of Cultural Security*, edited by Y. Watanabe. Cheltenham, UK: Edward Elgar, 2018.

———. *Surveillance in the Time of Insecurity.* New Brunswick, NJ: Rutgers University Press, 2010.

———, ed. *Surveillance and Security.* New York: Routledge, 2006.

Muller, Benjamin. "Travelers, Borders, Dangers: Locating the Political at the Biometric Border." In *Politics at the Airport*, edited by Mark B. Salter. Minneapolis: University of Minnesota Press, 2008.

Nakashima, Ellen, and Barton Gellman. "As Encryption Spreads, U.S. Worries About Access to Data for Investigations." *Washington Post,* August 10, 2015.https://www.washingtonpost.com/world/national-security/as-encryption-spreads-us-worries-about-access-to-data-for-investigations/2015/04/10/7c1c7518-d401-11e4-a62f-ee745911a4ff_story.html.

Nayar, Pramod. *Citizenship and Identity in the Age of Surveillance.* Delhi: Cambridge University Press, 2016.

Packard, Vance. *The Hidden Persuaders.* New York: Longmans, Greene, 1957.

Pallitto, Robert. "Bargaining with the Machine." *Surveillance and Society* 11, no. 1/2 (2013): 4–17.

———. "Irresistible Bargains: Navigating the Surveillance Society." *First Monday* 23, no. 2 (February 2018).

———. "Technology, Tradeoffs, and Freedom as Depicted in Postmodern Fiction." *Journal of American Culture* 40, no. 4 (2017): 399–413.

———, and Josiah Heyman. "Theorizing Cross-Border Mobility: Surveillance, Security, and Identity." *Surveillance and Society* 5, no. 3 (2008): 315–333.

Pasquale, Frank. "The Algorithmic Self." *Hedgehog Review* 17, no. 1(Spring 2015).

———. "From Territorial to Functional Sovereignty: The Case of Amazon." *Law and Political Economy* (blog). https://lpeblog.org/.

Rorty, Richard. *Contingency, Irony, and Solidarity.* New York: Cambridge University Press, 1989.

Rose, Nikolas. "Government and Control." *British Journal of Criminology* 40 (2000): 321–339.

Rosen, Christine. "The Machine and the Ghost." *New Republic,* July 11, 2012. https://newrepublic.com/article/104874/rosen-verbeek-technology-morality-intelligence.

Rotenberg, Marc, Julia Horwitz, and Jeramie Scott, eds. *Privacy in the Modern Age: The Search for Solutions.* New York: New Press, 2015.

Rousseau, Jean-Jacques. *The Social Contract and Later Political Writings,* 2nd ed. Edited and translated by Victor Gourevich. Cambridge: Cambridge University Press, 2019.

Sadowski, Jathan, and Frank Pasquale. "The Spectrum of Control: A Social Theory of the Smart City." *First Monday* 20, no. 7 (July 2015): 1–22.

Sagoff, Mark. *The Economy of the Earth: Philosophy, Law, and the Environment.* Cambridge: Cambridge University Press, 1988.

Salter, Mark B. "The Global Visa Regime and the Political Technologies of the Individual Self: Bodies, Biopolitics, Borders." *Alternatives* 31, no. 2 (2006): 167–189.

———. *Politics at the Airport.* Minneapolis: University of Minnesota Press, 2008.

Sandel, Michael. *What Money Can't Buy: The Moral limits of Markets.* New York: Farrar, Straus & Giroux, 2013.

Schneier, Bruce. "Fear and Convenience." In *Privacy in the Modern Age: The Search for Solutions,* edited by Marc Rotenberg, Julia Horwitz, and Jeramie Scott. New York: New Press, 2015.

Siemens. *The Business Case for Smart Cities* (Executive Summary). 2019. https://eu-smartcities.eu/sites/default/files/2017-09/BMF_Business%20Case%20for%20SC.pdf.

Singer, Joseph. "Something Important in Humanity." *Harvard Civil Rights-Civil Liberties Law Review* 37, no. 1 (2002): 103–130.

Solove, Daniel, and Danielle Citron. "Risk and Anxiety: A Theory of Data Breach Harms." *Texas Law Review* 96 (2018):737–786.

Sundeen, Mark. *The Man Who Quit Money.* New York: Riverhead, 2011.

Sunstein, Cass. "Terrorism and Probability Neglect." *Journal of Risk and Uncertainty* 26 (2003):121–136.

———. *Why Nudge? The Politics of Liberal Paternalism.* New Haven, CT: Yale University Press, 2015.

Srnicek, Nick. *Platform Capitalism.* Cambridge, MA: Polity, 2017.

Taussig, Michael. *Defacement: Public Secrecy and the Labor of the Negative.* Palo Alto, CA: Stanford University Press, 1999.

Thielman, Sam. "Silk Road Operator Ross Ulbricht Sentenced to Life in Prison," *Guardian,* May 29, 2015. https://www.theguardian.com/technology/2015/may/29/silk-road-ross-ulbricht-sentenced.

Valle, Victor. *City of Industry: Genealogies of Power in Southern California.* New Brunswick, NJ: Rutgers University Press, 2009.

Verbeek, Peter-Paul. *Moralizing Technology: Understanding and Designing the Morality of Things.* Chicago: University of Chicago Press, 2011.

Verchick, Robert. *Facing Catastrophe: Environmental Action for a Post-Katrina World.* Cambridge, MA: Harvard University Press, 2010.

Weaver, William, and Robert Pallitto. "State Secrets and Executive Power." *Political Science Quarterly* 120, no. 10 (2005): 85–112.

Weil, Simone. *Oppression and Liberty.* New York: Routledge, 2001.

Weinstein, Aviv, and Michel Lejoyeux. "Internet Addiction or Excessive Internet Use." *American Journal of Drug and Alcohol Abuse* 36, no. 5 (2010): 277–283.

Werbach, Kevin. "Trust But Verify: Why Blockchain Needs the Law." *Berkeley Technology Law Journal* 33 (2018): 487–550.

Werbach, Kevin, and Nicholas Cornell. "Contracts Ex Machina." *Duke Law Journal* 67 (2017): 313–382.

Winner, Langdon. *The Whale and the Reactor: A Search for Limits in an Age of High Technology.* Chicago: University of Chicago Press, 1986.

Wolff, Josephine. "The Most Shocking Thing about Encrypted Email Being Vulnerable Is That Anyone Still Uses Encrypted Email." *Slate,* May 15, 2018. https://slate.com/technology/2018/05/pgp-and-s-mime-are-vulnerable-but-also-no-one-used-them-anyway.html.

Wood, D., and S. Graham. "Permeable Boundaries in the Software-Sorted Society: Surveillance and the Differentiation of Mobility." In *Mobile Technologies of the City,* edited by M. Sheller and J. Urry. London: Routledge, 2006.

Wood, David Murakami. "Vanishing Surveillance: Securityscapes and Ambient Government." March 20, 2014. http://pactac.net/2014/03/vanishing-surveillance-securityscapes-and-ambient-government/.

INDEX